When the Land Meets the Sea

An ACUA and SHA Series

Series editors

Annalies Corbin, PAST Foundation, Columbus, OH, USA
J.W. Joseph, New South Associates, Inc., Stone Mountain, GA, USA

More information about this series at http://www.springer.com/series/8370

Alicia Caporaso
Editor

Formation Processes of Maritime Archaeological Landscapes

 Springer

Editor
Alicia Caporaso
Covington, LA
USA

ISSN 1869-6783
When the Land Meets the Sea
ISBN 978-3-319-48786-1 (hardcover)
ISBN 978-3-319-93181-4 (softcover 2018)
DOI 10.1007/978-3-319-48787-8

ISSN 1869-6791 (electronic)

ISBN 978-3-319-48787-8 (eBook)

Library of Congress Control Number: 2016955518

This Springer imprint is published by Springer Nature
The registered company is Springer International Publishing AG
The registered company address is: Gewerbestrasse 11, 6330 Cham, Switzerland

For Rachel

Acknowledgments

The genesis of this book began with the organization of a symposium on the formation of maritime sites and landscapes at the 2013 conference for the Society for American Archaeology at the University of Leicester, UK. I would like to thank Joyce Steinmetz for her assistance in organizing the session and Paul Johnston for synthesizing the main themes of the session as discussant. Each author who has contributed his/her/their chapter to this volume worked tirelessly to help me shepherd this book through the process of creation, editing, and production. I am very grateful to Annalies Corbin and the SHA publication committee as well as Teresa Krauss at Springer for supporting this project. Lastly, I must thank Ben Ford, my SHA Co-Publications Associate Editor, for allowing me to use his experience as the editor of the second volume in this series, to make the process so much easier than it may have been, and for his extensive, thoughtful comments on each chapter and the overall book.

Contents

1 Introduction 1
Alicia Caporaso

**2 A Dynamic Processual Maritime Archaeological
Landscape Formation Model** 7
Alicia Caporaso

**3 Mapping the Coastal Frontier: Shipwrecks and
the Cultural Landscape of the Early Republic Littoral** 31
Jamin Wells

**4 Collaboration, Collision, and (Re)Conciliation:
Indigenous Participation in Australia's Maritime Industry—
A Case Study from Point Pearce/Burgiyana, South Australia** 53
Madeline Fowler and Lester-Irabinna Rigney

**5 The Formation of a West African Maritime Seascape:
Atlantic Trade, Shipwrecks, and Formation Processes
on the Coast of Ghana** 79
Rachel Horlings and Gregory Cook

**6 Environment and Agency in the Formation
of the Eastern Ship Graveyard in the Central Basin
at Thonis-Heracleion, Egypt** 113
Damian Robinson, Franck Goddio and David Fabre

**7 Tsunami and Salvage: The Archaeological Landscape
of the Beeswax Wreck, Oregon, USA** 141
Scott S. Williams, Mitch Marken and Curt D. Peterson

**8 Coastal Erosion and Archaeological Site Formation
Processes on Santa Rosa Island, California** 163
Christopher S. Jazwa

**9   Formation Processes of Maritime Archaeological Sites
      in Guadeloupe (French West Indies): A First Approach** 189
      Jean-Sébastien Guibert, Christian Stouvenot and Frédéric Leroy

10   Conclusions/Discussion 211
      James P. Delgado

Index 219

Editor and Contributors

About the Editor

Dr. Alicia Caporaso has been a professional archaeologist for 12 years as an instructor of Archaeology, Cultural Anthropology, and Oceanography, and as a federal archaeologist with the National Park Service and the Bureau of Ocean Energy Management. Her publications and book reviews have appeared in The ACUA Underwater Proceedings, Le Journal, The Nebraska Anthropologist, The Northern Mariner, South Dakota History, and Technical Briefs in Historical Archaeology.

Contributors

Alicia Caporaso Bureau of Ocean Energy Management, New Orleans, LA, USA

Gregory Cook Department of Anthropology, University of West Florida, Pensacola, FL, USA

James P. Delgado Maritime Heritage, Office of National Marine Sanctuaries, National Oceanic and Atmospheric Administration, Silver Spring, MD, USA

David Fabre European Institute for Underwater Archaeology, Paris, France

Madeline Fowler Flinders University, Adelaide, SA, Australia

Franck Goddio European Institute for Underwater Archaeology, Paris, France

Jean-Sébastien Guibert AIHP-GÉODE EA 929 Université des Antilles, Schoelcher Cedex, France

Rachel Horlings Syracuse University, Syracuse, NY, USA

Christopher S. Jazwa Department of Anthropology, University of Nevada, Reno, NV, USA

Frédéric Leroy Département des Recherches Archéologiques Subaquatiques et Sous-Marines Direction Générale des Patrimoines, Ministère de la Culture et de la Communication, UMR 5140 (Archéologie des Sociétés Méditerranéennes), Marseille, France

Mitch Marken ESA, Palm Springs, CA, USA

Curt D. Peterson Geology Department, Portland State University, Portland, OR, USA

Lester-Irabinna Rigney University of South Australia, Adelaide, SA, Australia

Damian Robinson Oxford Centre for Maritime Archaeology, Institute of Archaeology, University of Oxford, Oxford, UK

Christian Stouvenot ArchAm (Archéologie des Amériques), Maison Archéologie & Ethnologie René-Ginouvès, CNRS UMR 8096, Nanterre Cedex, France; Service Régional de L'Archéologie de la Guadeloupe 28, Basse-Terre, Guadeloupe, France

Jamin Wells Department of History, University of West Florida, Pensacola, FL, USA

Scott S. Williams Turnwater, WA, USA

Chapter 1
Introduction

Alicia Caporaso

Between 2000 and 2012, the journal *Landscapes* published a series of essays written by British geographers, prehistorians, and artists entitled "What Landscape Means to Me." For most, approaching the subject evoked memories of exploration of the 'natural' and relict world as a child, the scope of which expanded with age and mobility. They describe how knowledge of a place leads to a sense of ownership, of experience, and of discovery. The recognition of this world as landscape required access and sensory revelation. Fundamentally, it is the antithesis of wilderness; it requires human intervention in its formation (Purslow 2006). It is "land which has been shaped" or domesticated (Fleming 2006). In all but one essay (O'Keeffe 2009), however, landscape is fully equated with rurality; it connects urban spaces, but does not include them. Archaeological sites and historic structures are incorporated into these landscapes only if the cultures that created them are gone or the activities that produced them are no longer present, and their remains are subsequently consumed by the environment. Several essays also invoke the concept of landscape decay (e.g., Fowler 2012), that these landscapes were fully formed *in the past*, and therefore can only now deteriorate. They all also require the presence of land—earth and vegetation—and access to the landscape in the present.

The descriptions of landscape presented in the *Landscapes* essays represent the most common archetype of the concept of landscape in popular, contemporary Western culture. *Maritime* cultural and/or archaeological landscapes (among others) incorporate so much more than these bucolic, agrarian paysages. What this signals to us as anthropologists, archaeologists, and historians is that the concept of landscape is mutable; its use in academic research requires explicit definition for particular studies, and, as will be apparent in the chapters that follow, landscape formation analysis can be approached in a variety of ways.

A. Caporaso (✉)
Bureau of Ocean Energy Management, 1201 Elmwood Park Blvd,
New Orleans, LA, USA
e-mail: Alicia.Caporaso@boem.gov; aliciacaporaso@yahoo.com

© Springer International Publishing AG 2017
A. Caporaso (ed.), *Formation Processes of Maritime Archaeological Landscapes*,
When the Land Meets the Sea, DOI 10.1007/978-3-319-48787-8_1

Ideas of landscape are traditionally used in maritime contexts in primarily four ways: the landscape of maritime economy; inundated formally lived surfaces; the setting of generalized coastal life; and the physical and interpretive management of archaeological resources (Firth 1997). For a true maritime landscape to exist, the society living in it must be attuned in some way to a body or bodies of water. Water must form a significant preoccupation of the society. The mere presence of the water does not define maritimity (Westerdahl 1998; Firth 1997). Rönnby (2007) has identified three structures, which he argues may be considered maritime *longue durées* (Braudel 1972) that may be used to define maritimity, as most maritime landscapes have them in common: exploitation of marine resources; communication over water; and the mental presence of the sea. Arnshav (2014:5, 8) adds to this list waste disposal into the sea as a way to get rid of discard and/or to create new land (infill). She notes that many cultures have developed a mythos of the sea in which it works to wash away anything considered dangerous, dirty, or physically and/or morally contaminating. I suggest that an additional, perhaps more modern (i.e., in the past three centuries) maritime *durée* is that of health, recreation, and aesthetic value, the latter including the commodification of the seaward viewshed.

In general, landscape incorporates a combination of referents, physical constructs or phenomena capable of being sensed, and signifiers, the sense and recognition/description/interpretation of the referent. Common referents in maritime archaeological landscapes may include purposeful coastal infrastructure such as docks, wharves, warehouses, etc.; punctuated inclusions such as shipwrecks; natural phenomena such as topography or the smell of sea air (Westerdahl 1992); and ephemeral phenomena such as seasonal lake ice (Ford 2011). They may be temporally restricted into "daymarks" (Parker 2001:36), components such as built or natural coastal ranging objects, for example church spires or mountain peaks, and "nightmarks" such as lighthouses. They also may be ever present but highly variable and sometimes unpredictable, for example currents, wind, and tides (Cooney 2003:324).

Common signifiers found in identified landscapes include myths, folklore, oral history, and codified phenomena such as the development of toponyms (Duncan 2011:268) or mapping standards (Crofts 2002). In maritime landscapes toponyms may be derived from several related sources such as important boathouses, harbors, major vessels, ship shelters, underwater barriers, shipwrecks, etc. (Stylegar and Grimm 2003). For example, Little Constance Bay in coastal Louisiana is named after the 1766 wreck of *El Nuevo Constante*, which occurred in the vicinity. The toponym helped archaeologists locate the remains of the shipwreck (Pearson and Hoffman 1995). Toponyms may also endure beyond memory of the original function, their true origin replaced by other explanations (Mägi 2008:86).

When the referent and signifier components endure temporally and spatially, a cultural discourse is transmitted and can evolve over successive generations (Layton and Ucko 1999). The patterning of these referents and signifiers, as with other forms of archaeological phenomena, can become residual, observable, and capable of being empirically studied in the landscape, primarily because they provide evidence of repetitious actions, and the patterning is visible at many spatial scales (Darvill 1999). The environment or natural landscape is therefore not neutral, but is a formidable agent in the formation and production of culture and can serve as its repository, literally, and figuratively (Harris 1999:434–436).

This landscape-tied patterning can also be transported to places other than those in which they were created. For example, the development of cognitive landscapes with physical environmental referents at the location at which a society or culture initially develops allows for the recognition of a place-type and a means of negotiating a newly accessed place in a colonial context, whether the location has been previously settled, with an indigenously developed social landscape, or not. In other words, the formation and negotiation of a new landscape is informed by the past experiences of its participants elsewhere.

Duncan (2011:271) has identified that for Pacific Island indigenous communities, cultural identity is codified within, and thus works to form the cultural landscape, in the names of places and stories told about them. History and belief is "read" by physically moving through the landscape. This harkens back to the childhood formative experiences described by the British authors of the "What Landscape Means to Me" essays. Their explorations codified, even if by invented, personal means, their belonging to their identified landscapes. Landscapes as identified, bounded, and in many cases, formally managed, real phenomena work to perpetuate these ties.

Maritime landscapes, when combined with historical information and statistics, are maps of human behavior in the long term. Approaching maritime archaeology from a landscape perspective reveals patterns of human behavior that can be empirically studied to reveal the constantly evolving process of negotiation within society and with the natural world. Westerdahl (2011) identifies five aspects of the maritime cultural landscape that may be the focus of archaeological research: the economic landscape; the resource landscape; the transport landscape; the cognitive landscape; and the ritual landscape. These represent the behavioral or cultural aspects of the maritime landscape. They may be shaped and/or constrained by the physical, geographical components of the landscape, by the land, water, or both, but they are not inherent in them.

Individuals, subgroups, or even entire communities that reside in or move through a geographical place may not identify the presence of or belonging to a particular landscape. They may recognize the landscape and adhere to its cultural constructs, but only have access to part of it. This is exceptionally true of maritime landscapes, where only a part of the community—sailors, fishermen, etc.,—directly accesses the sea. Individuals may participate in overlapping or interlinked cultural landscapes within the same geographical space, for example a maritime landscape, gendered landscape, or class-restricted landscape. There may even be cultural landscapes that some do not recognize at all as present or disregard as irrelevant.

Maritime landscapes may also be nested dependent on a spatial reference point with regard to a participant or group. At sea, if the spatial reference point is on a ship, the ship functions as if it is land, albeit a highly restricted geographical space, within which participants act. Behavior may not change regardless of the location of the ship. Other spatial reference points of great import in a maritime landscape to a sailor on a ship are land, water, and even sky. Participants do not transition from one reference point to another; all may be concurrently relevant. The combination of the effects of these spatial references informs behavior.

Research into the anthropogenic and taphonomic processes that affect the formation of maritime archaeological resources has grown significantly over the past decade in both theory and the analysis of specific sites and associated material culture. The addition of interdisciplinary inquiry, investigative techniques, and analytical modeling, from fields such as engineering, oceanography, and marine biology have increased our ability to trace the unique pathways through which archaeological sites progress from initial deposition to the present, yet they can also link individual sites in an integrated social, environmental, and geographic maritime landscape.

As we can elicit and study the processes by which maritime archaeological sites are formed, so can we too the processes that form and perpetuate maritime social and archaeological landscapes from the past to our interpretation of them in the present. An easy example of the formation of a maritime archaeological landscape is the submergence of an active cultural and/or archaeological landscape that initially formed subaerially. A submerged archaeological landscape may or may not have formed prior to inundation, and its pre-inundation landscape may not have been a maritime landscape. For example, it can be argued that Port Royal, Jamaica was always a part of a maritime cultural landscape both pre- and post-inundation. Conversely, a site inundated by a dam, perhaps well away from the affected river, may in no way be oriented toward maritimity before it was drowned. In these two examples, the archaeological remains of the former, inundated as the result of a catastrophic incident, would be completely different from a planned event, such as the latter. Preparation for a purposeful inundation might include the removal of all personal material goods and commercial structures identified as having inherent or economic value, materials considered toxic or dangerous, and the population, including even the relocation of interred burials.

However, an archaeological site or landscape that was purposefully inundated may transform into a maritime landscape through both cultural and natural formation processes. The site may become a natural hazard to shipping, changing sailing practices, which may result in contemporary mitigation such as removal. It may become a fish haven, acting as a reef, which may create or alter commercial and recreational fishing practices. It may become a site for recreational diving. Natural transforms such as biological colonization, chemical reactions with seawater, and physical processes such as storm surge and wave action will impact the site(s) as they would any other submerged component of a maritime landscape. Additionally, archaeological practice also defines a landscape as maritime in the way in which we approach its identification and spatial delineation, investigation, and management (e.g., Tuddenham 2010).

The Formation of Maritime Archaeological Landscapes presents a global perspective of current research in maritime archaeological landscape formation processes. Subject maritime landscapes include both those that develop contemporaneously with real and/or potential component archaeological sites as well as modern landscapes that develop through evolving interaction with select archaeological sites. The organization of the chapters follows an arc from theoretical discourse to application of landscape formation interpretation. Detailed case studies will look at the development of functional paradigms for incorporating archaeological, sociocultural, and historical data into the formation of maritime landscapes (Caporaso; Wells); interdisciplinary scientific study of archaeological site components to elicit information on maritime behavior not visible in traditional artifact analyses (Horlings and Cook; Robinson et al.; Williams et al.); and modern uses of maritime archaeological landscapes that remake and redefine meaning in archaeological discourse (Jazwa and Fowler and Rigney).

In addition to "classically" considered submerged material cultural and geography, or those that can be accessed by traditional underwater methodology, these case studies include less-often considered sites and landscapes that require archaeologists to, for example, use geophysical marine survey equipment to characterize extensive areas of the seafloor or go above the water's surface to access maritime archaeological resources that have received less scholarly attention. Also included are landscapes that initially formed on land, but are now submerged, which require a combination of both terrestrial and underwater techniques to tease out the circumstances of their combined formation processes. Together, these studies show that the formation of the maritime archaeological landscape and changes in maritime activity are irreducibly diachronically and spatially linked and the co-evolution of landscape and behavior is evident in the archaeological record.

References

Arnshav, Mirja. 2014. The freedom of the seas: Untapping the archaeological potential of marine debris. *Journal of Maritime Archaeology* 9: 1–25.

Braudel, Fernand. 1972. *The mediterranean and the mediterranean world in the age of Phillip II.* London: Collins.

Cooney, Gabriel. 2003. Introduction: Seeing land from the sea. *World Archaeology* 35(3): 323–328.

Crofts, Roger. 2002. What landscape means to me. *Landscapes* 2: 103–106.

Darvill, Timothy. 1999. The historic environment, historic landscapes, and space-time-action models in landscape archaeology. In *The archaeology and anthropology of landscape: Shaping your landscape*, ed. Peter J. Ucko, and Robert Layton. London, England: Routledge.

Duncan, Brad. 2011. "What do you want to catch?": Exploring the maritime cultural landscapes of the queenscliff fishing community. In *The archaeology of maritime landscapes*, ed. Ben Ford, 267–289. New York: Springer.

Firth, Antony. 1997. *Three facets of maritime archaeology: Society, landscape, and critique.* http://avebury.arch.soton.ac.uk/Research/Firth. Accessed 5 March 2009.

Fleming, Andrew. 2006. What landscape means to me. *Landscapes* 1: 122–126.

Ford, Ben. 2011. The shoreline as a bridge, not a boundary: Cognitive maritime landscapes of lake Ontario. In *The archaeology of maritime landscapes*, ed. Ben Ford, 63–80. New York: Springer.

Fowler, Peter. 2012. What landscape means to me. *Landscapes* 1: 57–69.

Harris, Dianne. 1999. The postmodernization of landscape: A critical historiography. *The Journal of the Society of Architectural Historians* 58(3): 434–443.

Layton, Robert, and Peter J. Ucko. 1999. Introduction: Gazing on the landscape and encountering the environment. In *The archaeology and anthropology of landscape: Shaping your landscape*, ed. Peter J. Ucko, and Robert Layton. London, England: Routledge.

Mägi, Marika. 2008. Facing the sea. In S. Lilja (Ed.), *Leva vid Östersjöns kust. En antologi om naturförutsätningar och resursutnyttjande pa bada sidorav Östersjön ca 800–1800* [*Life on the Baltic Coast: An anthology of natural conditions and resource utilization on both sides of the Baltic Sea*]. Flemingsberg, Sweden: Södertörns University.

O'Keeffe, Tadhg. 2009. What landscape means to me. *Landscapes* 1: 123–130.

Parker, A.J. 2001. Maritime landscapes. *Landscapes* 1: 22–41.

Pearson, Charles E., and Paul E. Hoffman. 1995. *The last voyage of El Nuevo constante: The wreck and recovery of an eighteenth-century spanish ship off the louisiana coast*. Baton Rouge, Louisiana: Louisiana State University Press.

Purslow, Richard. 2006. What landscape means to me. *Landscapes* 2: 105–115.

Rönnby, Johan. 2007. Maritime durées: Long-term structures in a coastal landscape. *Journal of Maritime Archaeology* 2: 65–82.

Stylegar, Frans-Arne, and Oliver Grimm. 2003. Place-names as evidence for ancient maritime culture in norway. *Årbok Norsk Sjøfartsmuseum* 2002: 79–115.

Tuddenham, D.B. 2010. Maritime cultural landscapes, maritimity and quasi objects. *Journal of Maritime Archaeology* 5: 5–16.

Westerdahl, Christer. 1992. The maritime cultural landscape. *The International Journal of Nautical Archaeology* 21: 5–14.

Westerdahl, Christer. 1998. *The maritime cultural landscape: On the concept of the traditional zones of transport geography*. Institute of Archaeology and Ethnology: University of Copenhagen, Copenhagen, Denmark.

Westerdahl, Christer. 2011. Conclusion: The maritime cultural landscape revisited. In *The archaeology of maritime landscapes*, ed. Ben Ford, 267–289. New York: Springer.

Chapter 2
A Dynamic Processual Maritime Archaeological Landscape Formation Model

Alicia Caporaso

Introduction

Human behavior and its resultant material remains exist on a physical and cultural landscape and cannot be separated from it. Studying archaeological sites within the landscape reveals patterns of human behavior that can only be identified within that context. At the same time, it is imperative to consider the environmental regime and human interactions with and within it; otherwise, it is impossible to fully understand the maritime landscape and associated human behaviors.

The shipwreck and its archaeological site are components of the narrative of human behavior on the maritime archaeological landscape. The natural environment in turn constrains and informs human behavior and plays a large and important role in the development of maritime culture and the maritime landscape. The processes by which this occurs can be studied through analysis of the archaeological record. This chapter integrates the components of the maritime landscape with the understanding of the archaeological and historic records as well as oceanographic processes to develop a model that takes into account not only shipwrecks but all archaeological remains in a region. While the maritime cultural and archaeological landscape incorporates all behaviors, beliefs, activities, and physical and social constructs oriented socially and spatially toward maritimity, this chapter specifically focuses on the landscape of maritime disasters—how the cultural and physical environment influences wrecking events—to show the proposed model's efficacy. This is subsequently illustrated by the example of spatial and temporal patterning of shipwreck events in and around Thunder Bay in northern Lake Huron in the nineteenth and early twentieth centuries.

A. Caporaso (✉)
Bureau of Ocean Energy Management, 1201 Elmwood Park Blvd,
New Orleans, LA 70123, USA
e-mail: Alicia.Caporaso@boem.gov; aliciacaporaso@yahoo.com

© Springer International Publishing AG 2017
A. Caporaso (ed.), *Formation Processes of Maritime Archaeological Landscapes*,
When the Land Meets the Sea, DOI 10.1007/978-3-319-48787-8_2

Maritime Cultural and Archaeological Landscapes

A true archaeological landscape cannot be fully defined materially, but has to be understood as a both physical and cognitive social, spatial, and temporal construct of what is physically present and what those who live within and those who study it perceive. As it will be shown, this does not mean that the landscape cannot be studied scientifically; rather it informs a more robust empirical understanding of human behavior and its relationship with the natural world.

Traditionally, archaeologists have viewed the landscape in one of two ways: as a physical phenomenon of human construction focusing on the human–land relationship in economic terms, or as a subject, reconstructing snapshots of historical elements (Darvill 1999:105). Darvill (1999:108–110) states that these approaches necessarily overemphasize the built landscape that can be experienced visually. They also assume that the landscape is essentially stable in at least the short term. He recommends considering the landscape as a socially imposed conception of space, time, and social action on what is perceived to be the natural and social world. The partitioning of space and time can be physical or cognitive, defined through attributed meaning. Social action is different from general behavior as it must be collectively intentional. In other words, society structures landscape.

Westerdahl (1998) identifies two fundamentally important socially constructed physical components of the maritime landscape: transport zones and maritime technology. Transport zones are enduring or traditional zones of transport geography. They requires community consent and cognitive recognition for their existence. They exist in physical space yet their parameters of use are structured socially. Two parts to their understanding are long-term perspective where transport zones have associated direction (vector) attributes and their cultural, environmental, and technological restrictions of use such as transport techniques, climatic adaptation, seasonal variations, anthropogenic factors such as technology use-skill, etc. For example, a zone may not exist during the period of time in which it is iced over, but exist with the appearance of icebreakers. Shipping accidents often occur when activity operates outside of transport zone boundaries or when the attributes of transport zones are altered, such as during violent storms. Zone identification can vary depending on spatial scale of analysis; they can split and reform in response to outside forces; they can be temporary or periodic; and the loci of transfer between zones can be dangerous, often the location of several wrecking events, the consequences of non-uniform change on the landscape.

A second category of maritime transportation zones can be patterned onto the above physical transportation zones: the duality of danger versus safe zones (Duncan 2004). Both danger and safe zones have physical and cognitive definitions, and identification varies according to the same parameters as the physical transport zones; however, what sets them apart from the fundamental zone types is the fact that their physical and cognitive identifications can often be in opposition to one another.

The significance of maritime technology is that it is assumed to be adapted to the transport zones in which it is being operated. It is also dependent on peripheral social factors such as risk recognition and responsive behavior. The combination of the two components allows for a landscape approach to maritime behavior incorporating both the social and natural components of the landscape.

A useful paradigm for treating the landscape in the formation of a conceptual regional site formation process-based model is to treat the maritime landscape as a "nonlinear dynamical system whose evolution is governed by abrupt transitions" (McGlade 1999). This does not preclude the presence and efficacy of non-abrupt transitions, but necessitates that they are not the primary mode of or condition rates of social and physical change. McGlade (1999) refers to this as a human ecodynamic approach, concerned with the dynamics of human-modified landscapes from a long-term perspective. This therefore is useful for the consideration of maritime landscapes and associated transport zones. The human–environment relationship involves the co-evolution of socio-historical and environmental processes and their intersection in time and space producing the socio-natural system as an analytical framework (McGlade 1999:462).

Sociocultural evolution is spurred by "positive" feedback, which produces temporary and/or permanent destabilizing social effects that push society through unstable transitions (McGlade 1999:464). The socio-natural system can be conceptualized as a framework of stored system energy. System efficiency is realized when energy enters the system as pulses (Odum 2007). Shipping accidents and accident mitigation, the invention of new shipping technology, and the discovery of a valuable natural resource can be considered systemic energetic pulses. Within an ecodynamic system, a small change in one variable can have catastrophic effects on the system as a whole. The social response therefore can be considered as a form of self-reorganization rather than mere adaptation to the effects of positive feedback (McGlade 1999:464). Often however, affective change is partial, with parts of society able to withstand change more than others.

Increased diversity of the temporal and spatial scales of phenomena within the landscape increases lag in social evolution, the results of which can cause disruptive sociocultural and natural dynamics (Gould 1983; McGlade 1999:465). Because in the short term, these system responses are nonlinear, it is necessary to define patterns in long-term behavior. In the long-term, discontinuous transitions that result in patterns of social activity and environmental events in nonequilibrium are normal (McGlade 1999:465).

Therefore, understanding patterns of behavior in the long term is the only really effective way to understand the behavioral components of maritime disasters and associated shipwreck distribution, and a landscape of maritime disasters is a very useful place to compile relevant data. Having a model within which to collate, interpret and communicate the data—and give due attention to oceanographic factors that influence the data—makes it comparable to similar research conducted in other areas, so that landscapes and individual archaeological sites can be connected to greater landscape processes.

The Evolution of Maritime Archaeological Site Formation Models

Formation processes of the archaeological record have been a concern of terrestrial archaeologists for much of the second half of the twentieth century, and constitute one of the tenets of the New Archaeology paradigm developed in the 1960s (Binford 1983). Hypotheses about formation processes were developed in the late 1960s–1970s through the application of the concept of entropy to archaeological sites; i.e., that potential site-derived information degrades over time. During the 1970s, the idea of transformation processes recognized that there is discontinuity between human activities, artifact deposition, and preservation and archaeological recovery or the creation of sampling bias. Previous work has shown that formation processes: transforms, sites, and regions formally, spatially, quantitatively, and relationally, can create distortion and artifact patterns unrelated to past human behaviors, but exhibit regularities that can be studied and expressed statistically (Schiffer 1987:9–11).

Work in model development for the understanding of shipwreck formation processes has lagged its terrestrial counterpart, but has continued to slowly evolve since Muckelroy first proposed his "evolution of a shipwreck" in 1978. All discourse on shipwreck formation processes understand that the environment and other natural factors contribute to the creation and, to some degree, modification of the archaeological record. Formation theory provides structure for applying information that is derived from the site (O'Shea 2002:10). Gould (1983:18) points out that many maritime and nautical archaeologists state that they implicitly include all the steps of "archaeological reasoning" in their research, including site formation, but that there is a need to be explicit in how the rules of science are applied to explanations of past human behavior. This "implicit" inclusion leads to inconsistent application of environmental factors in research and contributes to "the illusion of site uniqueness" (O'Shea 2002:3). Though recent research on specific shipwreck sites has at times been able to pinpoint specific natural events that have contributed to the wrecking of individual vessels or groups of vessels, scientists are just beginning to understand how the environment affects submerged cultural materials and vice versa (Lenihan et al. 1981; Forsythe et al. 2000; Jordan 2003). At both the local and regional levels, this is necessary "to develop a reasonable perspective in the rational utilization of the [archaeological] resource base" (Murphy 1983:80–81).

Even the latest models of site formation processes, those that include the oceanographic or limnological aspects of the formation process, fail to provide a thorough understanding of archaeological formation and preservation when one considers a maritime landscape. They oversimplify or ignore the movement of archaeological materials. This is not to say that this problem has yet to be addressed. Regional studies of shipwrecked materials, which take into account environmental factors, have been carried out both in the United States and abroad (Wheeler 2002; O'Shea 2002, 2004). These studies, however, do not explicitly delineate a model that that can be applied generally to most or all maritime land-scapes. They also only lightly touch upon the probability that environmental regimes may have quite specific preservation potentials which carries with it regional archaeological management significance (Wheeler 2002:1151).

To fully understand the proposed landscape formation model, it is necessary to derive its evolution through previous modeling attempts. Muckelroy was the first to explicitly put forth that shipwreck phenomena contain common features (Fig. 2.1). This implies that when evidence can be ascertained and tested on sites where historical evidence is present, it can also be applied where historical evidence is lacking. Therefore, archaeological evidence is inherently homogenous with at least some degree of cohesion and the assemblage can be approached as a system defined by the characteristics of a ship which may have gone through a series of trans-formations through time (1978:157–159). Muckelroy's shipwreck evolution model interprets the site formation process as a closed system with only the ship as an input. There are extractive (salvage, disintegration, dissolution, etc.) and scrambling devices, which include the wrecking process itself and seabed movement.

Based on review of studies that attempt to measure the quality of archaeological remains through parallel biological and geomorphological marine studies,

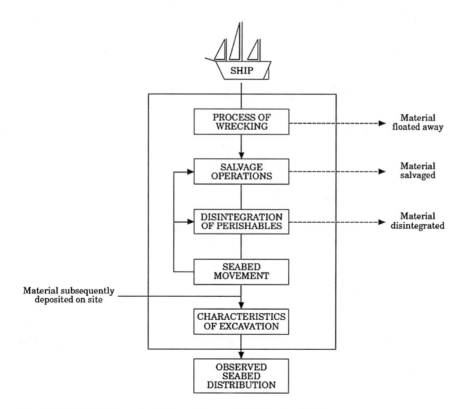

Fig. 2.1 The first shipwreck site formation model (Muckelroy 1978)

Muckelroy (1978) characterizes what he considers to be relative environmental attributes that contribute to material extraction and scrambling. He determined that the geomorphology or nature of the seabed deposit is the most important control for determining site scrambling and cohesion (survival) of archaeological remains underwater. The most deterministic forcing factors are: maximum offshore fetch, sea horizon or open water, average slope of the seabed, recent underwater topography, and coarse versus fine sediment as deposit matrix. Less relevant attributes are tidal stream speed and depth of site.

Muckelroy's attempt to classify well-known shipwreck sites according to cohesion, presence, and amount of extant archaeological material with observed environmental impacts is descriptive and not causal. He states that it is not possible to predict the likelihood that remains will be found in a location known to have been a wrecking site, and that the model neither addresses variability between sites of similar geomorphology, nor does it looks at the wrecking process as a single event (1978:165).

This formation model is rather simplistic, as it does not consider inputs that are themselves defined in the present as archaeological such as floral and faunal attraction to the site (as habitat) or post-"shipwreck" anthropogenic input such as salvage process debris, memento deposition, net snags, etc. It also does not allow for extraction due to non-"floating away at time of wrecking" means. There are other extracting filters besides salvage and disintegration once the site has reached "stabilization" such as storm surges, currents, waves, ice movement, etc., especially in lacustrine environments.

While in general there is little cause to dispute Muckelroy's deterministic factors affecting material extraction and site scrambling, it is erroneous to believe that they can neither predict the probability that archaeological remains will be found in a given location nor that they can address site variability. The probability that remains might be found in a given location is inherent in Muckelroy's process of wrecking if the deterministic factors are taken into account. Both site location and intersite variability can be addressed if the model is expanded to include the material once it has been extracted from the primary archaeological site, or in other words, if archaeological material is recognized to remain a tangible part of the landscape. In addition, O'Shea (2002:8) notes that the use of "scrambling device" as a term for material movement implies a randomization or pattern diminishing effect (entropy), which is inaccurate.

Nearly a decade after Muckelroy proposed his site formation model, Schiffer (1987) more precisely defined how to characterize site transformation processes by dividing them into two categories, cultural and natural, termed in turn c-transforms and n-transforms, and breaking down process effect components to three levels: the artifact, the site, and the region. While Schiffer was considering terrestrial archaeological deposits and does not appear to intend the work to be considered a model per se, it can be used to define both physical and spatial inputs into an underwater formation process model. Most importantly, it is understood that sites are open systems and, therefore, one should include inputs other than the initial deposit and materials once they have left the immediate site area (1987:151).

C-transforms are defined as the processes of human behavior that affect or transform artifacts after their initial period of use in a given activity and are responsible for retaining items in a systematic context to form the historic and archaeological records as well as for any post-depositional modification. N-transforms are any natural processes that effect archaeological deposits by deterioration, decay, alteration, or other modification and can add environmental material to the site. Unlike c-transforms, they are to some degree continuous. Resulting transformation/modification for both types of transforms are both regular causally and consequently making the processes and their effects predictable and thus able to be statistically modeled (Schiffer 1987:7, 21–22, 143).

Schiffer (1987:199) also shows how when approaching the archaeological record, c- and n-transforms can be invariably linked. Non-cultural processes will affect behavior that potentially causes c-transforms to occur. For example, environmental factors might keep sailors from venturing into certain places due to historically understood geological, physical, or other concerns (shoals, cross-currents, whales, etc.).

N-transforms must also be taken into account when approaching an archaeological landscape. They can affect both site visibility and accessibility. They can also bias survey and sampling regimes, for example, sedimentary processes may variably expose or cover some or all of an archaeological site rendering its identification in a side-scan sonar survey dependent on the sedimentary conditions on the day that the survey takes place. Identifying formation processes in the archaeological record implies that they occurred. It is necessary; therefore, to be explicit as to their effects and without extensive analysis it may be impossible to separate the archaeological remains from them. This is especially true when one considers that there can be much variability in the effect of transforms (Schiffer 1987:265–267, 302).

In any case, it is important to identify formation processes before behavioral and environmental inferences are made so they can be filtered from the anthropological phenomena of interest (Schiffer 1987:303). Gould (1990:21, 53–54) terms these as first-order and second-order variables. First-order variables are the constraints of the environment (n-transforms) as well as anthropological limitations, such as the state of technology at a given time. Second-order variables are the "human factor" or specific behaviors (c-transforms) that will aid in a better understanding within a cultural-historical context (i.e., desired anthropological information). In other words, formation process controls must be ordered. The need for second-order variables defines Gould's "Operational Theory", which assumes cultural uniformitarianism and a form of middle-range theory (contemporary and historical) to derive c-transforms that can be applied to and/or filtered from the archaeological remains (1990:49, 55). Examples of operational theory are the "One More Voyage Hypothesis" (Murphy 1983) and "Technological Trend Innovation". The latter is the perpetual improvement, including increased complexity and cost, of a traditional industrial system that over time is rendered increasingly obsolete. At least one segment of the social hierarchy has a stake in the perpetual production of the system (Gould 1990:170–189).

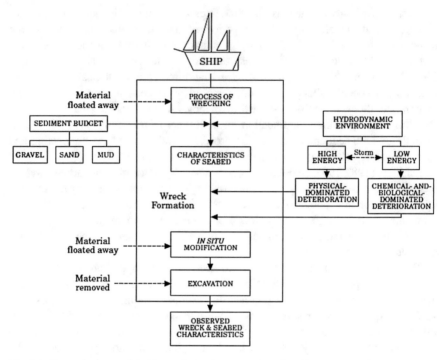

Fig. 2.2 Dynamic processual shipwreck site formation model (Ward et al. 1999)

Ward et al.'s (1999:561–563) processual site formation model was the first to incorporate dynamic natural transforms into the core of the model (Fig. 2.2). It is predictive, process-oriented, and scale-independent. Shipwreck degradation is characterized by the sum of direct and indirect affective environmental processes plotted against the local sediment budget (considered a first-order control) over depositional history. The environmental processes are physical or hydrodynamic, biological, and chemical. It is clear that most, if not all, of these processes are interrelated and cannot be considered apart from one another. This model is excellent for characterizing site formation processes within the context of an individual archaeological site.

The sediment budget is defined as the rate of net supply or removal of different types and sizes of sediment grains to the site area. Ward et al. (1999:564–565) recognize the link between sediment budget and hydrodynamic forcing; however, consider it separately because one does not assume the other bidirectionally. The sediment budget influences the extent of development of reduction-oxidation zones within the sediment. For many sites, it may be possible to examine sediment profiles to determine the history of the sediment budget.

The nature of the hydrodynamic environment is variable in time. Physical effects have greater impact in high-energy environments and biological and chemical effects have greater impact in low-energy environments. The site can transition to and from high-energy environments to low-energy environments, and at any stage, material can be lost. There are an infinite number of different process paths a site could progress through to reach the "present" state (Ward et al. 1999:565, 568).

Ward et al. (1999) approach the visual interpretation of the model as a revision of Muckelroy's (1978) flowchart. This flow chart adds the sediment budget and the hydrodynamic environment as inputs into the process of pre-stability site formation; therefore, the wrecking process is no longer a single unidirectional path toward site stability. Obviously, different site types degrade or are affected in dissimilar ways, but it is clear that these factors have a greater influence on the modification of the site than if one only considers the ambient "steady-state" environment (Ward et al. 1999:564). This equilibrium environmental characterization can be used to normalize the forcing factors when comparing sites and/or loci.

The most recent dynamic site formation model assumes Ward et al.'s (1999) natural transformational process, expanding it to include the range of cultural processes, before, during, and after a shipwreck and the long-term relationships between people and shipwreck sites (Gibbs 2006:4–5) (Fig. 2.3). Gibbs (2006:7) argues that cultural transforms should be structured, not as pre-depositional, depositional, and post-depositional, but around the nature of the event and the sequence and range of potential responses at each stage of the event.

Using Leach's five major stages of a physical disaster, a shipping disaster follows the following processes (Gibbs 2006:7–13):

1. Pre-impact: the pre-impact stage comprises a period of recognized potential threat and a period of warning in which evidence indicates that an accident is likely to occur. A threat may be real, manufactured, or imagined (Duncan 2004:15) and it may or may not be understood. During the warning period, mitigation can be physical and/or spiritual and successful mitigation can result in an arrest in accident progression.

2. Impact: the impact stage is the moment of a disaster event through the realization that the event has occurred and mitigation must take place. Disaster studies have shown consistent trends within groups during the impact stage with only a small proportion of the group able to respond immediately and effectively. The remainder is often bewildered or behaves inappropriately. Mitigation may include jettisoning cargo or fixtures, patching a leak, or intentionally grounding a vessel.

3. Recoil: the recoil stage commences when the immediate threat to life has receded or that the primary disaster event has been survived. This does not mean that involved individuals are out of danger. It is possible that a vessel can be successfully mitigated out of both the impact and recoil stages resulting in no shipwreck archaeological site. Other event-related materials might be retained to form the archaeological record (e.g., flotsam, jetsam, etc.).

4. Rescue: the rescue stage commences when the person or group involved in the disaster has been removed from danger. Often, this is where many of the first documentary accounts of the event are generated such as in life-saving station logs or rescue vessel logs.

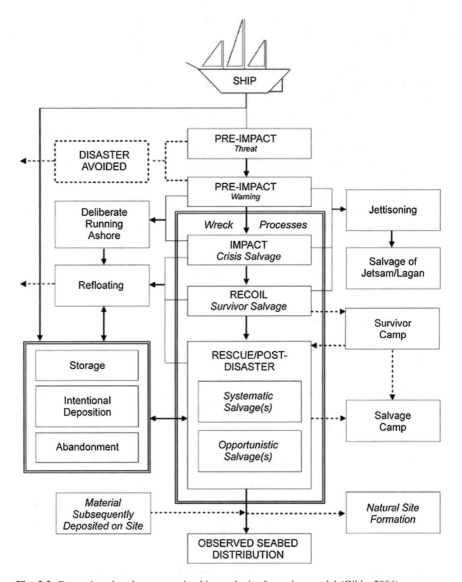

Fig. 2.3 Dynamic cultural processes in shipwreck site formation model (Gibbs 2006)

5. Post-trauma: post-trauma is the medium- and long-term response to the event. Most primary documentary accounts of an event are produced at this time. This can include insurance reports, newspaper articles, etc.

Gibbs (2006:14–15) also recognizes the importance of salvage as a key formation process of archaeological sites and rightly recognizes the variability in the methods and effects of salvage in different disaster stages. Salvage can begin during the recoil stage and continue long after the disaster participants are no longer actors in the life of the shipwreck. Salvage is variable over time and is dependent on site accessibility, the time and effort required to salvage, the perceived benefits versus cost, and the legality of the endeavor. Opportunistic and organized salvage can occur in several cycles and in either order. This also applies to mobilized wreckage such as cargo, other materials, and even corpses that wash ashore.

Gibbs arranges the pathways of his processual model by modifying Muckelroy's flowchart to include explicitly defined c-transforms present in terms of the five stages of the shipwreck event. In other words, it follows the process of the associated human activity. Ward et al.'s (1999) natural transformation model is the continuation of the dynamic formation processes that affect the site apart from human interference and therefore does not overlap with the cultural transforms.

Though it acknowledges archaeological materials distributed off-site through human activity, the model is designed to specifically address the formation of a single archaeological site; it only implicitly assumes the presence of removed material elsewhere. This in itself is not a failing of the model, but rather the model assumes that what is of interest is the formation of the immediate site location of the shipwrecking event and that associated materials would be included when investigating the particulars of associated activities. Gibb's (2006) model does, however, only include purposeful cultural transforms; it does not include inadvertent or incidental human activity that affects shipwreck sites such as channel dredging or snagging towed gear.

While the Ward et al. (1999) and Gibbs (2006) models appear to be excellent for a rigid site definition, they do not allow at all for site parameter flexibility. The site must derive from a single event and is therefore strictly locational. It is not useful for characterizing an archaeological landscape and all the archaeological materials contained within. Additionally, the preceding models in general do not allow for the inclusion of many types of archaeological materials located with the maritime archaeological landscape. This includes wreckage that has broken away from shipwreck sites, either in the process of wrecking, or after the archaeological site has initially formed. Archaeological material can derive also from other, non-ship types of maritime transportation and infrastructure such as log rafts, fishing, and other non-transportation activities. All of this cultural material is an integral part of the maritime archaeological landscape.

A Maritime Archaeological Landscape Formation Model

To create a useful maritime archaeological landscape formation model, archaeological space and time must be analyzed in three dimensions, including the surface and water column in addition to the sea floor (Fig. 2.4). There are three levels or stages of analysis within this system upon which variables can act. In terms of a vessel these are: (1) that a vessel will wreck or become irrecoverable in a given location at the surface; (2) that wreckage will arrive at a given location; and (3) and

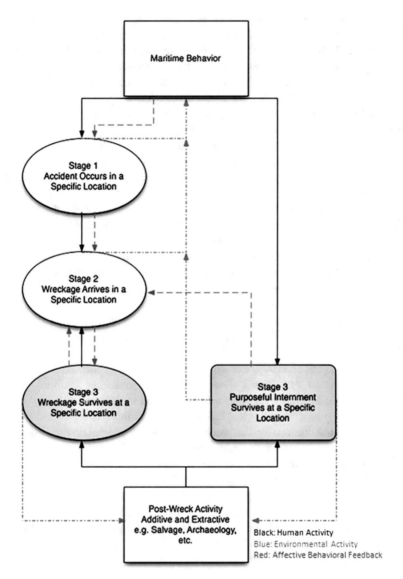

Fig. 2.4 Dynamic processual maritime archaeological landscape formation model

that wreckage will survive at a given location. Each shipwreck or wreckage must go through each of these stages of transformation. Additionally, once reaching the third stage, mobile wreckage may become re-entrained within the system, due to c- and/or n-transforms, and continuously cycle through stages 2 and 3. While in stage 3, the Ward et al. (1999) model is applicable for all submerged archaeological remains.

The model is generalized in order to be applied to any definable maritime landscape. Formation transforms are defined as environmental inputs that can be characterized or measured in space. They can be both variable and non-variable and exist at different scales. These transforms can be measured directly in the environment, derived from historic data, or inferred from historic accounts. Depending on the area and scale studied, different transforms will have more influence than others.

In order to approach the maritime landscape as a human ecodynamic system, it is impossible to separate anthropogenic and natural phenomena (McGlade 1995:359) as with the models of Ward et al. (1999) and Gibbs (2006). The former treats c-transforms as an arbitrary initial input and the latter does not incorporate n-transforms into the sociocultural model except to assume the forcing of behavioral constraints over time. Additionally, neither allows for positive feedback from landscape events over time to affect the continual cycling of landscape formation processes. The landscape is not merely a sum of the events that take place within it.

To combine anthropogenic and environmental processes into a single landscape formation model, it is necessary to treat the landscape as an irreducible socio-natural system (McGlade 1995:359). Additionally, though previous formation cycles inform human behavior, individual shipwreck and other internment events are essentially mutually exclusive. It is a nonlinear system where behavior cannot be reduced to a mathematical algorithm. What is needed is a model that acts as an abstract dialogic resource that can carry multiple analytical arguments through a variety of model scenarios and various temporal and spatial scales (McGlade 1995:361). This approach allows for the combination of different types of data including descriptive, deductive, and interpretive datasets.

McGlade (1995:361–366, 384) has developed an organizational structure in which this approach is possible. A framework is required in which empirical data are situated within an interpretive as opposed to a deductive frame of reference in order to facilitate an interrogative dialogue between qualitative and quantitative data. Instead of a model as a representation of real-world phenomena, the model becomes a dialogic resource constructed around the potential interaction between model sets within which multiple possible arguments can be formed. Each problem set or inquiry requires appropriate submodels to address different aspects of observed phenomena within different boundary domains. Instead of a single predictive model, inquiry leads to a series of potential evolutionary pathways to which a system is prone. The model allows for selective component access to address specific problems related to the dynamics of the system. The modeling process provides an "experimental arena" within which different interpretations are possible hypotheses that can be tested.

Unlike all previous models all human behavior associated with maritime activities is incorporated with the landscape formation model, because they fundamentally take place within it. This includes behaviors that are not directly included in the wrecking process such as successful maritime voyages and modern salvage activities.

Each stage in the model can be construed as a "black box", which represents an infinite number of potential c- and n-transforms that are bounded by the unique conditions present at a given time and that evolve over time. Except for "Maritime Behavior", which can be construed as being continuous throughout all or limited to a part of the landscape, every stage within the model is associated with a particular location on the landscape. As people and materials move between stages, this location may stay the same or change.

Once materials arrive at Stage 3, on the sea floor, an archaeological site is formed. There are two pathways through which this can occur, through an accidental internment (Stage 3a) or through a purposeful internment (e.g., pound net stakes, wharf pilings, etc.; Stage 3b). On the landscape, these sites are contingent on the moment of observation. While some archaeological sites persist through their initial formation to the present, others may have a finite lifespan controlled through time-dependent c- and n-transforms.

Both cultural and natural forces drive the system through the three stages. These forces are time dependent. For physical changes on the landscape, these can be either unidirectional or bidirectional. The bidirectional forces create cycles within the system. The dominant cyclical physical process is the entrainment of archaeological materials from an established site into the water column, essentially returning it to Stage 2, where it will then be deposited elsewhere creating a new site. Environmental forces involved might include storm-induced currents and waves or ice scouring, and human forces might include purposeful activities such as dredging and dumping of spoil.

An accumulation of input into any stage may split and continue the formation model at different locations ultimately forming unique archaeological sites within the landscape. For example, cargo may be jettisoned during accident recoil; a portion of the crew may leave the vessel in a lifeboat; wreckage may differentially disperse on the surface; or a portion of material may become entrained into the water column and be carried from an archaeological site.

What makes this model truly a landscape formation model, as opposed to an archaeological site population model with mutually exclusive site formations taking place intrasite, is the creation of positive feedback that drives the continuous flux of human behavior, which in turn drives the entire system as it affects the primary input: maritime behavior. This feedback can initiate from several stages, may not be the same at different stages, and may affect the overall system differentially. Both the feedback itself and the effects of the feedback are contemporarily unpredictable. It may be ignored, misunderstood, perverted, or dismissed, and disparate groups or individuals might use and respond to it differently.

Additionally, because maritime behavior and environmental forces occur as parts of the landscape as a whole, areas within the landscape that do not contain archaeological materials retain their importance within the system and cannot be discounted or ignored. Just as dynamic behavioral and physical processes are deterministic factors in the creation and presence of archaeological sites, they equally inform the lack of sites and any given place.

The continuous cycling of the dynamic formation of the landscape creates a system that can absorb the effect of force inputs. Systemic steady state is not synchronous between the landscape and processes occurring intrasite. Perturbations within the system at any stage may affect or may not affect a particular location or site but always affect the landscape. Because the formation of the landscape is tied to both time and place, different parts of the model can be accessed to address specific questions posed to it. Understanding every possible input into the model is not required for it to function as a dialogic resource for inquiry.

Formation of the Maritime Archaeological Landscape of Thunder Bay, Lake Huron: Decade of Loss

Thunder Bay is located in northwestern Lake Huron near the city of Alpena, Michigan. The 4300 square miles of sanctuary hold at least 100 known and identified shipwrecks and perhaps another 100 unidentified shipwrecks, at various depths ranging from 0 to over 100 ft. Most of the known shipwreck sites and many of the unknown shipwreck sites, derive from familiar, well-known ships and events. Dating primarily between 1860 and 1930, historic records reveal their names, the identity of their crew, the form and construction of the vessels, their cargo, and when, where, why, and how they wrecked.

The ~ 70 years of shipping disaster represented by these shipwrecks indicate that safety at sea was a real pervasive concern for those tied to the lake for their livelihoods. The differences in the circumstances of these shipwrecks allow us to tease out behaviors tied to the landscape. For example, Thunder Bay was considered the only safe harbor during storms along Lake Huron's northwestern coast. Many of the ships that wrecked at Thunder Bay foundered or stranded while seeking shelter along the southern shore of North Point peninsula. Additionally, pilot books from the late nineteenth and early twentieth centuries promoted the lee passage between Thunder Bay Island and North Point as safe harbor for ships in peril. This passage, however, is characterized by dangerous reefs, which require careful knowledgeable piloting, a difficult chore in calm weather let alone in heavy seas. Pecoraro (2007) has shown that this lee passage actively functioned as a ship trap that served to concentrate maritime accidents into a condensed area.

Through the use of this maritime archaeological landscape formation model, patterns and trends in commercial shipping and associated human behavior become readily apparent in the submerged archaeological record. As a whole, the spatial patterning of shipwrecks in the region is not random. Nearest neighbor analysis of all historic shipwrecks (Fig. 2.5), even when the position of any shipwreck only

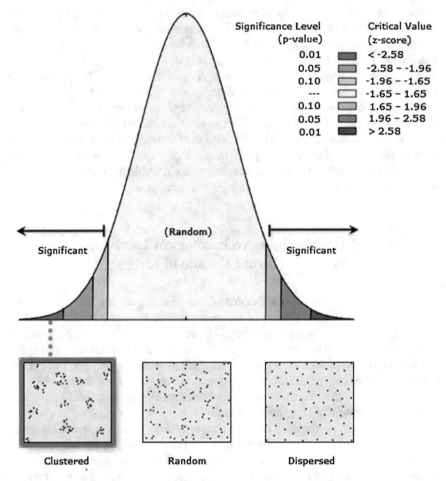

Fig. 2.5 Nearest neighbor analysis of shipwrecks in the vicinity of Thunder Bay

known through the historic record is estimated, indicates that there is a less than 1%
likelihood that the pattern is the result of random chance, further supporting the
model theory that formation processes do not begin with an archaeological site's
initial deposition on the lake floor. In the context of the model, it becomes easy to
explore how the social conditions of the period between 1830 and 1930 informed
the maritime behavior that best explains the historical and spatial distribution of
archaeological materials associated with primary shipwreck sites. This discussion
considers patterns associated with decade of loss (Figs. 2.6 and 2.7).

The total number of known shipwrecks in the vicinity of Thunder Bay in the
1830s through the 1850s is relatively low due to the small number of vessels on the
upper Great Lakes at the time, little commercial activity around Lake Superior, and
possibly because of limited recording of shipwreck accidents, though as they would

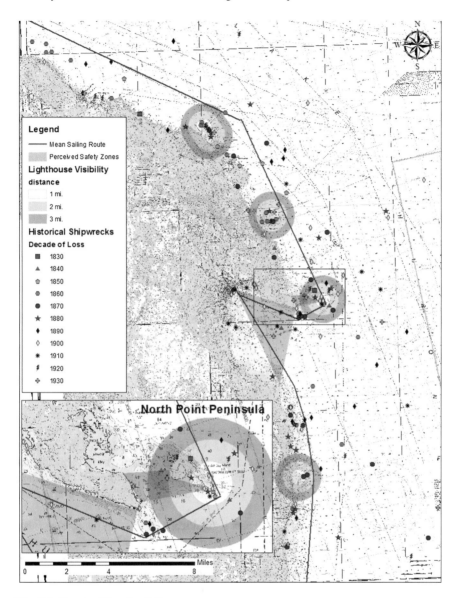

Fig. 2.6 Decade of loss (spatial)

have been rare and costly, any severe accident would likely have been noted. Because steam-powered vessels were expensive, fewer risks would have been taken with their use; therefore, it is unsurprising that only four wrecked during these decades.

A lack of spatial patterning of shipwrecks during these decades indicates unfamiliarity with the coastline, as the coast had yet to be surveyed and pilot books

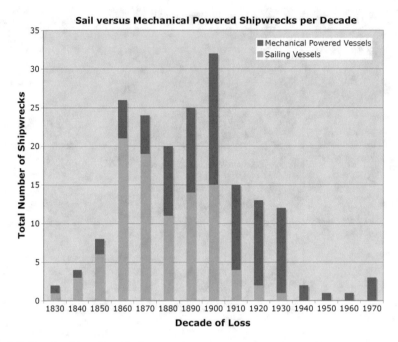

Fig. 2.7 Decade of loss (graphical)

were unavailable. All shipwrecks before 1860, however, took place within or immediately adjacent to the coastal transport zone either in shoals or within the historic shipping lane. It appears at this time that the corridor between Thunder Bay and Sugar Islands and North Point and the lee side of Presque Isle were becoming recognized as a perceived safety zones for imperiled vessels as several wrecks occurred in these locations. This informal designation was probably the primary motivation for the construction of the regions two earliest lighthouses, one at each of these locations.

The period of 1860 through 1900 saw a significant increase in the total number of shipwrecks in the vicinity of Thunder Bay with sailing vessels making up at least half of the wrecks for every decade except 1900. This indicates that as the ratio of new steam-powered to sailing vessels increased, sailing vessels became increasingly marginalized as a profitable method of shipping and variable risks would have been taken by sailing and steam-powered vessels owners. It also explains the split in the rates of growth, with sailing vessels being much higher, of the average ages of shipwrecks of these two propulsion systems throughout this period. The percentage of sailing vessels that shipwrecked compared with the total fleet also increased throughout this period, while the percentage of steam-powered vessels that wrecked remained stable. The latter is not unexpected, as overall, fleet size grew with tonnage availability even as cargo capacity became increasingly larger.

An interesting spatial patterning of shipwrecks occurred between 1860 and 1900. The 1860s saw significant clustering of shipwrecks at three primary locations: at Forty-Mile Point, Presque Isle, and Middle Island. All of these locations were nodes of sailing direction change. Forty-Mile Point was a new navigational node, beginning in 1860s with the development of copper mining, for vessels sailing to and from Lake Superior. This is also the area where vessels sailing from Lake Michigan and Lake Superior merge to form a single southbound corridor. Nearly every shipwreck at this node sank due to collision indicating unfamiliarity with the route and a lack of recognized formal sailing directions.

The perceived danger zone at Presque Isle was mitigated by the construction of a new lighthouse. The 1870s saw no apparent overall clustering. This changed in the 1880s with apparent clustering at Thunder Bay Island and Presque Isle and also north and south of the new lighthouse at Sturgeon Point. As few wrecks had occurred previously at Sturgeon Point, this indicates that the lighthouse became a perceived safety zone on the maritime landscape at this time. Also, new pilot books in the 1860s and 1870s had formalized the North Point corridor as a perceived safety zone on the landscape. Again, there was little clustering in the 1890s, then, again, clustering in the 1900s around Thunder Bay Island.

This cycle of clustering over a 50-year period likely represents a behavioral reaction on a decadal scale of the maritime community to perceived dangers on the landscape. When apparent clusters of shipwrecks occur in a recognized period of time, both the perceived and real risks are mitigated, through coastal survey, the distribution of pilot books, and the erection of lighthouses, and in the latter half of this period the institution of life-saving stations. While accidents do occur at these locations afterward, the pattern of shipwreck location becomes much more diffuse. Over time, the institutionalization of their presence and their associated risk mitigation, coupled with changes in maritime technology and general maritime transport and commercial shipping conditions, forces the cycle to begin anew with new patterns of clustering that fit the new maritime landscape conditions.

The number of shipwrecks in total fell precipitously between 1910 and 1930. Few wrecks of sailing vessels illustrate their essentially complete marginalization in commercial shipping. All new commercial vessels built during this time were iron, the majority steel freighters, yet most of the shipwrecks of steam-powered or other mechanized vessels at this time were old wooden-hulled boats. Only a small percentage carried bulk commodities and those that did carried relatively non-valuable cargos compared with iron ore, therefore it is likely that these vessels were for the most part competing in regional trading activities. A preponderance of tugs indicates that many accidents occurred in a local context. Throughout this period, the spatial distribution of shipwrecks continued to contract in general toward Thunder Bay.

Hot-spot analysis of shipwrecks of sailing vessels in the region reveals some interesting patterns (Fig. 2.8). Hot-spot analysis measures the standard deviation in spatial clustering of an attribute relative to other individual points, point clusters, and attribute clusters. A hot-spot will have points of closely clustered attributes spatially restricted from other points or clusters. A cold spot will have clusters of points but with mixed attributes. Lastly, a neutral spot will have either clusters of an

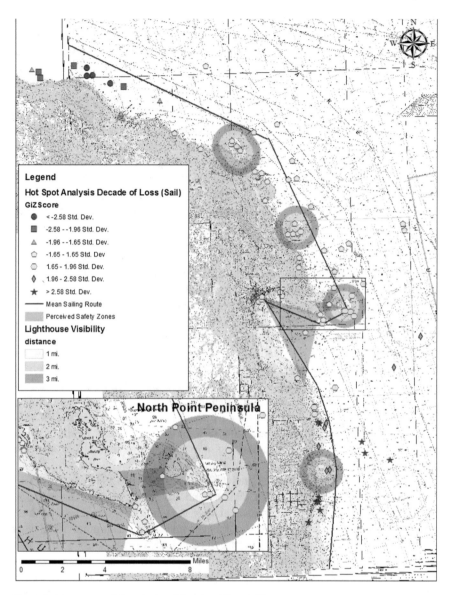

Fig. 2.8 Hot spot spatial analysis of decade of loss (sail)

attribute in close proximity to clusters of another attribute or a diffuse collection of points. In the vicinity of Thunder Bay, shipwrecks northwest of Presque Isle form hot spots from the 1860s and the 1890s. Moving southward along the coast, there are neither hot nor cold spots between Presque Isle and North Point indicated that clusters of shipwrecks in any given decade do occur; however, they are in close proximity to other clusters of shipwrecks and shipwrecks that are diffuse on the landscape. This is not unexpected in an area with a large overall number of ship-wrecks compared with the rest of the region. There are cool spots located around Thunder Bay Island and North Point indicating clusters of wrecks from several decades. South of South Point, shipwrecks from several decades are highly clustered in proximity to one another. There are no hot or cold spots with regards to steam-powered vessels.

Conclusion

In general in Western society, oceans and lakes have been considered as in the realm of the chaotic, each event mutually exclusive. The outcome of the voyage was thought of as inherently random. Human behavior on the maritime landscape and the environmental variables, which inform and constrain it, however, are not chaotic and random. Approaching maritime archaeology from a landscape perspective reveals patterns of human behavior that can be empirically studied to reveal the constantly evolving process of negotiation within society and with the natural world.

While methodological approaches to the study of the physical landscape are generally quantitative, it is impossible to combine attributes of the social cognitive landscape with it that can fit into these analytical paradigms. Human behavior that works to produce the archaeological record in any given instance is not fully predictable; however, the result of these actions is measurable in both the historical and archaeological record. The lack of predictability explains why existing models that provide a systemic framework for the analysis of archaeological formation processes begin with the identified results of behavior, the creation of a shipwreck or other submerged archaeological site. The most robust of these models, Ward et al. (1999), is excellent for the analysis of dynamic formation processes on unique archaeological sites that are inherently linked with specific locations on the land-scape. This is adequate for the analysis of intrasite archaeological formation processes and their associated physical expression.

These, models, however, discount socially constrained human behavior that works to create the opportunity for the archaeological sites to form. They also decouple the result of the action/accident from affecting both future maritime activity and the evolving understanding of the past. Additionally, these models are fundamentally tied to place. When mobilized wreckage is entrained within the physical system and moved elsewhere or is salvaged from the archaeological site, it is eliminated from consideration in the models.

The maritime archaeological landscape retains all archaeological materials within and associated with it as well as the human behavior and, perhaps more importantly, the contemporary and historical understanding of its results and the associated social responses to it. These are, of course, variable, as they are dependent on interest, environmental and technical knowledge, and economic and social power. They also evolve variably over time. Tangible evidence for it may be found in the archaeological record or derived from the historical record; however, this information must be interpreted leading to essentially qualitative analyses of the maritime landscape.

The maritime archaeological landscape formation model allows for the incorporation of both quantitative and qualitative analyses, conducted at variable temporal and spatial scales, into a single cogent interpretive framework. The model functions as a dialogic resource or arena in which subsections of the framework can be accessed in order to address specific scientific questions or to formulate hypotheses that can be addressed through nested modes of inquiry. The model allows for an infinite number of nested models of any type within all of the enclosed realms of activity present in the maritime landscape. The power of this model to produce new hypotheses in which to pose to the archaeological record is well-illustrated in the above example of nineteenth and early twentieth century sailing in and around Thunder Bay in northern Lake Huron.

The proposed hypothesis is that participants in this maritime landscape recognized patterns in accidents related to commercial shipping on a decadal scale. Feedback packaged at a decadal periodicity was powerful enough to affect maritime behavior. By looking at the patterns of commercial shipwrecking activity within the maritime landscape by decade, it has been possible to reveal evolving patterns of maritime behavior over time. It is clear that individual accidents that could potentially result in shipwrecks were expected, and maritime behavior changed little specifically in response to them. What was required to fundamentally change maritime behavior was extra-normative discoveries, inventions, internal and external pressures, events, and accidents, the latter often involving shipwrecking.

In the classical anthropological sense, Muckelroy is correct in his overarching statement:

> The study of the wrecking process itself is of limited intrinsic significance, its importance lying rather in the link it provides between the remains investigated and the original vessel. Furthermore, the potential and limitations of our understanding of the latter by archaeological means ultimately defines the scope of the whole subdiscipline of maritime archaeology (1978:215).

This is not a comprehensive view for why we need to study archaeological site formation processes. With the recognized importance of being stewards managing the archaeological landscape, understanding the nature of the archaeological landscape and its associated maritime behavior is as important as determining the anthropological and historical information that can be derived from it.

References

Binford, Lewis R. 1983. In *Pursuit of the past: Decoding the archaeological record.* Berkley, California: University of California Press.
Darvill, Timothy. 1999. The historic environment, historic landscapes, and space-time-action models in landscape archaeology. In *The archaeology and anthropology of landscape: Shaping your landscape*, ed. Peter J. Ucko, and Robert Layton. London, England: Routledge.

Duncan, Brad. 2004. Risky business, the role of risk in shaping the maritime cultural landscape and shipwreck patterning: A case study application in the Gippsland Region, Victoria. *Bulletin of the Australasian Institute for Maritime Archaeology* 28: 11–24.

Forsythe, W., C. Breen, C. Callaghan and R. McConkey. 2000. Historic Storms and Shipwrecks in Ireland: A Preliminary Survey of Severe Synoptic Conditions as a Causal Factor in Underwater Archaeology. *International Journal of Nautical Archaeology* 29(2): 247–259.

Gibbs, Martin. 2006. Cultural site formation processes in maritime archaeology: Disaster response, salvage and muckelroy 30 years on. *International Journal of Nautical Archaeology* 35(1): 4–19.

Gould, Richard A. 1990. *Recovering the past*. Albuquerque, New Mexico: University of New Mexico Press.

Gould, Richard A. 1983. The archaeology of war: Wrecks of the Spanish Armada of 1588 and the Battle of Britain, 1940. In *Shipwreck anthropology*, edited by Richard Gould. Albuquerque, New Mexico: University of New Mexico Press.

Jordan, Brian A. 2003. Analysis of environmental conditions and types of biodeterioration affecting the preservation of archaeological wood at the kolding shipwreck site. Dissertation on file: University of Minnesota, St. Paul, Minnesota.

Lenihan, Daniel, J., Toni L. Carrell, Stephen Fosberg, Larry Murphy, Sandra L. Rayl, and John A. Ware. 1981. *The final report of the national reservoir inundation study*, vols. 1, 2. Santa Fe, New Mexico: United States Department of the Interior, National Park Service, Southwest Cultural Resources Center.

McGlade, James. 1995. An integrative multiscalar modelling framework for human ecodynamic research in the Vera Basin, south-east Spain. *L'Homme et la degredation de l'environement*, ed. Van der Leeuw S. E. Juan les Pins, APDCA.

McGlade, James. 1999. Archaeology and the evolution of cultural landscapes: Towards an interdisciplinary research agenda. In *The archaeology and anthropology of landscape: Shaping your landscape*, ed. Peter J. Ucko, and Robert Layton. London, England: Routledge.

Muckelroy, Keith. 1978. *Maritime archaeology*. London, England: Cambridge University Press.

Murphy, Larry. 1983. Shipwrecks and database for human behavioral studies. In *Shipwreck anthropology*, ed. Richard A. Gould. Albuquerque, New Mexico: University of New Mexico Press.

Odum, Howard T. 2007. *Environment, power, and society for the twenty-first century: The hierarchy of energy*. New York: Columbia University Press.

O'Shea, John M. 2002. The archaeology of scattered wreck sites: Formation processes and shallow water archaeology in western lake huron. *International Journal of Nautical Archaeology* 31(2): 211–227.

O'Shea, John M. 2004. The identification of shipwreck sites: A bayesian approach. *Journal of Archaeological Science* 31: 1533–1552.

Pecoraro, Tiffany A. 2007. Great lakes ship traps and salvage: A regional analysis of an archaeological phenomenon. Thesis on file: East Carolina University.

Schiffer, Michael B. 1987. *Formation processes of the archaeological record*. Albuquerque, New Mexico: University of New Mexico Press.

Ward, I.A.K., P. Larcombe, and P. Veth. 1999. A new process-based model for wreck site formation. *Journal of Archaeological Science* 26(5): 561–570.

Westerdahl, Christer. 1998. *The maritime cultural landscape: On the concept of the traditional zones of transport geography*. Copenhagen, Denmark: Institute of Archaeology and Ethnology, University of Copenhagen.

Wheeler, A.J. 2002. Environmental controls on shipwreck preservation: The Irish context. *Journal of Archaeological Science* 29: 1149–1159.

Chapter 3
Mapping the Coastal Frontier: Shipwrecks and the Cultural Landscape of the Early Republic Littoral

Jamin Wells

Introduction

Sad is the scene,–despair frowns 'mid the wreck; Hopeless, benumb'd, worn out, they strive in vain (*Philadelphia Repository and Weekly Register* 1802).

Yellow fever chased *Industry* out of Aux Cayes, Hispaniola, in late November 1801. Bound for Boston, Massachusetts, the two-masted schooner set out along a well-traveled route that skirted the Cuban coast before heading north through the Florida Straits to New England, paralleling the eastern shore of the fledgling United States for more than 1000 miles. Besides a handful of dispersed port cities, the American coast was a dark and desolate frontier, sparsely inhabited and isolated from the centers of American life. Captain Gideon Rae probably planned to avoid as much of this treacherous coast as possible by keeping *Industry* safely offshore in the Gulf Stream. The schooner's crew figured that they could, with luck, be home in time to celebrate Christmas. Their voyage did not go according to plan.

Industry's master and mate succumbed to yellow fever within a week of leaving Aux Cayes. For the short-handed crew of five that remained, the schooner's destination still lay over 2000 miles away. Fortunately, the route was well known for these young but experienced mariners. Down two sailors, the crew's daily routine of "hard, steady, and generally familiar work" navigating the two-masted craft

J. Wells (✉)
Department of History, University of West Florida, Bldg 50/Rm. 112,
11000 University Parkway, Pensacola, FL 32514, USA
e-mail: Jwells2@uwf.edu; jaminjwells@gmail.com

© Springer International Publishing AG 2017
A. Caporaso (ed.), *Formation Processes of Maritime Archaeological Landscapes*,
When the Land Meets the Sea, DOI 10.1007/978-3-319-48787-8_3

simply became more onerous. *Industry* languished off the Cuban coast for 3 weeks, accompanied by 30 or 40 other vessels that were also hindered by strong head-winds. Once in the Gulf Stream, however, the schooner cruised north along the American coast, reaching southern New England from "off Havana" in just 5 days (Lamson 1908:164–170).[1]

By Sunday, December 27, 1801, *Industry* was standing off the recently lit Montauk Point lighthouse, on the eastern end of Long Island, New York, when the wind veered to the west-northwest and "blew heavy." The schooner hove to—sails and rudder adjusted so it maintained its position—for 3 days before the weather moderated and it continued toward Boston. By Wednesday afternoon, another wind shift brought a blinding snowstorm from the east. For more than 10 hours, the short-handed crew sailed in the "heavy gale," struggling to keep *Industry* from being driven ashore on Cape Cod. They failed. A little after one-thirty in the morning on the last day of 1801, Zachary Lamson, acting master of *Industry*, saw "breakers alongside." Lamson, a curly haired teenager just shy of his 19th birthday, was already on his eighth voyage. This was his first shipwreck and he would remember it vividly decades later. The schooner struck the sandy coast of Cape Cod. The shock broke the foremast in half, sending it overboard. Massive breakers washed the deck. The sailors cut away the mainmast, lightening the schooner until it "beat over the shoal and drove up on the beach" (Lamson 1908:166–167).

Shortly after, the men jumped off *Industry*'s bowsprit, touching land for the first time in more than 5 weeks. They landed on a featureless stretch of snow-covered sand, seeking shelter from the storm behind a nearby hill. For 3 hours they walked in circles to "keep ourselves from freezing." As Lamson recalled: "At daylight we travelled to the south not knowing what part of the Cape we were on. Having walked about four miles and no prospect of relief, I concluded to return, as we were getting feeble." Only three made it back to *Industry*; the others had been left behind on the beach, too weak to continue. "On our arrival at the vessel," Lamson continued, "we found her keeling over on shore and the tide had left her, so we could get to her side, but we were so exhausted, we could only…tumble and crawl on our hands and knees to the cabin, as our clothes…were frozen and our strength gone."

The gale subsided as the shipwrecked crew slept. Half an hour before sunset, residents from the nearby towns of Chatham and Orleans discovered the stranded *Industry*. They acted quickly to revive the men in the cabin. A search party located the two who had been left on the beach "in a senseless state buried in part in the snow." A boat carried the survivors six miles to the nearest house where three women nursed them back to life. "The only inconvenience," Lamson reported, "was the loss of the skin from their faces and hands, and such parts of their bodies as were most exposed. My suffering was very bad … I could not walk for twelve days" (Lamson 1908:167–175).

---

[1] I have not been able to locate the logbook for the *Industry*. However, Lamson, *Autobiography*, 164–167 includes a fairly precise account. Schooner *Industry*'s crew included: Gideon Rae, master; John Stone, mate; Zachary Lamson (18 years old), John Low, Peter Woodbury, and John Porter (21 years old), seamen; Cato Gowing, cook (Lamson 1908:164–170).

Shipwrecks are processes that continue long after the rescue of survivors. Despite his injuries, Lamson took responsibility for the vessel and cargo, directing the salvage of *Industry* from the saddle of a horse. Successful salvage required swift action and he spent the month of January directing efforts to save what he could of the wreck—contacting the schooner's owners in Boston; hiring men to remove the cargo of coffee and "all the wreck worth saving;" protecting the salvaged goods from conspiring locals; and organizing the transportation of it all to Boston. No local officials oversaw the operation and only after the last salvaged remnants were loaded on a chartered schooner did a representative of *Industry*'s owners appear on the scene.[2] Satisfied with Lamson's efforts, he sent the 18-year-old sailor to Boston with the goods and settled outstanding accounts in Chatham and Orleans. What became of the salvaged cargo is unclear, but on February 3 the *Boston Gazette* published this advertisement: "Sales at Auction: THIS DAY, at No. 61 Long Wharf, for the benefit of the underwriters. A quantity of SAILS, standing and running RIGGING, CABLES, & c. saved from the wreck of the schooner *Industry*, lately stranded on Cape Cod."[3] Zachary Lamson went home to Beverly, Massachusetts tired but undaunted; he soon returned to sea (Lamson 1908:169–175).

Industry's wreck and salvage is representative of the innumerable shipwrecks that occurred in the early American littoral. Shipwrecks were not remarkable in early nineteenth-century America. Indeed, at least four other vessels wrecked on Cape Cod in January 1802, joining 25 others reported as wrecks along the American coast that month. There were likely more; systematic accounting of American coastal shipwrecks did not begin for another 70 years.[4] Newspapers at the beginning of the century grouped these everyday disasters in "shipping news" columns typically located near the end of each edition. The *Salem Gazette*'s account of the *Industry* wreck is representative: "… we are informed, that there were four vessels ashore on Cape Cod, 3 schooners and a sloop; the latter belonged to New Bedford … One of the schooners is owned in this town … the names of the other two schooners we have not heard" (*The Salem Register* 1802). One of those "other two" schooners was *Industry*, which for contemporaries was just another

---

[2]By law, the local sheriff would have taken responsibility of the wreck and cargo. Lamson makes no references to any local oversight. Nor does he mention Customs, which would have had an interest in the wreck. No relevant local or customs records survive.

[3]Apparently there were no takers—the same advertisement was published the following day. The second auction was presumably more successful; the ad was not published again.

[4]We do not know how many vessels actually wrecked in the early republic littoral. Cook (2013:30) offers a sound estimate that 4–5% of vessels were "castaway or lost at sea." Newspapers reported individual wrecks, but they did not discuss shipwrecks synoptically. Even the preeminent maritime information source, *Lloyd's List*, which had reported shipping casualties since 1741, did not systematically tabulate casualty returns until 1890. Further, no historian to my knowledge has undertaken the herculean task of tabulating turn-of-the-century shipwrecks. Reconciling the overlapping, inconsistent, and often contradictory accounts of marine disasters that do exist could still only provide a rough estimate.

wreck on the desolate, dangerous coastal frontier and an occurrence too common to warrant more than a passing reference.

Events surrounding early republic shipwrecks like *Industry* illustrate some of the characteristics of the American maritime cultural landscape before it was dramatically remade by regulation and domestication. At the beginning of the nineteenth century, the American littoral was an isolated, parochial, preindustrial space strewn with shipwrecks on the margins of a fledgling nation. The shipwrecks that defined the American littoral were subtly yet irrevocably altering this maritime cultural landscape. In the decades after Independence, coastal shipwrecks became one of the few reasons why outsiders trekked into this veritable frontier. Federal agents came to secure the duties that financed government operations, humanitarian organizations to succor the shipwrecked, and entrepreneurs to acquire and commodify the local knowledge that was essential for safe navigation. At the same time, newspaper editors, artists, authors, and other cultural producers brought tales of coastal shipwrecks into the homes and workplaces of Americans, familiarizing them with (if not actually domesticating) the wild coast and disaster through sensational narratives. Coastal shipwrecks, as this chapter demonstrates, are not only events, material remains, processes, or narratives that illuminate a maritime cultural landscape, but ones that defined and fundamentally shaped one.

The American Coastal Frontier

The American coast was a frontier when *Industry* wrecked in 1801. Few terms are as historiographically fraught as *frontier* but stripped of its post-1890s Turnerian associations, frontier aptly describes the littoral as turn-of-the-century Americans would have understood the term. Noah Webster defined "frontier" in 1806 as "a limit, boundary, border on another country, furthest settlements." It was a definition that would not substantively change in popular dictionaries for generations (Mood 1948:79). The Atlantic coast—*the* American coast at the beginning of the nineteenth century—served as both an unavoidable physical limit and a political border for the United States.[5] Passamaquoddy Bay marked its northern terminus, separating the state of Massachusetts (present-day Maine) from the English colony of Nova Scotia (present-day New Brunswick). To the south, the circuitous St. Mary's River divided Georgia and Spanish Florida. Between these borders stretched 1500 miles of towering cliffs, rock-strewn inlets, and barren sandy beaches. The political border lay three miles offshore, an invisible line drawn in a fluid environment. Since 1793 the United States had claimed exclusive jurisdiction to "the utmost range of a cannon ball," as Secretary of State Thomas Jefferson had

---

[5]Of course the east coast was not the only "American" coast, which also included the shores of Lake Erie and the Mississippi River. But at the opening of the nineteenth century, the Atlantic coast was not only the scene of the preponderance of waterborne commerce but also captured the public imagination in a way these other shores would not for decades.

explained at the time (Oberg and Looney 2008). With good reason did Joseph E. Worcester's 1860 *A Dictionary of the English Language* illustrate its definition of frontier with this explanation: "the best *frontier* is the sea" (Mood 1948:80).

Few people lived in the turn-of-the-century American littoral. Only an occasional fishing hamlet and handful of scattered ports punctuated an otherwise desolate shoreline.[6] The nation's largest cities in 1800—New York, Boston, Philadelphia, Baltimore, and Charleston—were situated on or adjacent to the coast. Yet natural barriers—rivers, bays, estuaries, salt marshes, and pine barrens—coupled with preindustrial transportation and communication networks concentrated these cities and other coastal settlements.[7] By 1799, for example, New York handled almost one-third of the nation's overseas trade and one-fourth of its coasting trade. Yet the port's densely populated, bustling waterfront hugged, "like a rock in a sock," the southern tip of an island still dominated by pastures, swamps, and marshland (Burrows and Wallace 1999:333–343; Stiles 2011:14). On the coast, the 103-foot octagonal Sandy Hook Lighthouse, constructed at the entrance of New York Bay in 1764, offered the only human-constructed navigation mark for more than 100 miles north or south of the port's primary entry point. New York's natural advantages—a large, protected deep-water harbor adjacent to a navigable river that flowed hundreds of miles inland—set it apart from the rest of the American littoral. In less-geographically favored places, settlers avoided the barren and unforgiving coast by founding villages and towns relatively far from the sea and closer to inland natural resources and transportation routes. Sites that would in the following decades hold a central place on the American maritime cultural landscape remained on its periphery at the beginning of the nineteenth century. The sea, for example, "held little attraction" for the founders of Long Branch, New Jersey—the future site of the country's preeminent nineteenth-century seaside resort. The initial settlement, located 1.5 miles from the coast, remained isolated from the small collection of fishing huts on the immediate coast until the second quarter of the nineteenth century (Writers' Project 1940:7–25). Indeed, while nearly two-thirds of the American population still lived within 50 miles of tidewater, its geographic center had shifted more than 40 miles west between the first and second censuses suggesting the degree to which settlers spurned the immediate shoreline (U.S. Census Bureau 2001).

Beyond the densely populated coastal region between New York and New Hampshire, the immediate shoreline was inhabited, on average, by fewer than six people per square mile—the definition of frontier adopted by the U.S. Census Bureau in 1874 and taken up by subsequent generations of demographers and

---

[6]Native Americans had been largely removed from the East Coast. By 1750, just a handful of "Indian Remnants" remained in the Greater New York littoral (Meinig 1986:208–212).

[7]American life and labor remained constrained by the world they lived in. This small-scale, adaptable, and thoroughly local world—what Lewis Mumford terms the *eotechnic age*—was powered by wind and water and built of wood and stone. Irregularity plagued the eotechnic age; its dependence on strong steady winds and the regular flow of water "limited the spread and universalization of this [eotechnic] economy" (Mumford 1934).

historians (Mood 1945:24–30).[8] According to the 1800 federal census, the majority of the coasts of Maine, New Jersey, Delaware, North Carolina, South Carolina, and Georgia as well as substantial sections of the Massachusetts, Rhode Island, and New York coastlines met the bureau's definition of frontier (U.S. Census Bureau 1898). County-level census data, however, actually obscures just how desolate the nation's eastern border was. After traveling through Long Island's relatively populous north shore, Yale University president Timothy Dwight concluded: "The inhabitants, though considerably numerous on paper, are yet in each township, to a great extent, scattered in a very thin dispersion" (Dwight 1823:274). He noted, however, the diversity of Long Island's coastal society, which was undergirded by farmers and fishermen but increasingly home to an array of specialists, including innkeepers, ferrymen, and settled ministers. Dwight was no unalloyed booster and he offered this succinct description of his nation's maritime culture landscape in another letter written during the first decade of the nineteenth century: "The American coast, as you know, is chiefly barren, and of course thinly inhabited" (Dwight 1823:70).

Human settlement was so sparse along the coast that the occasional house, windmill, and other human-made structures stood out against the natural landscape and became an indispensable aid to coastal navigation. Captain McCobb's mill, on the west side of Maine's Kennebec River, marked the farthest north a prudent mariner lacking local knowledge sailed without the assistance of a local pilot. Captain Henderlon's red house and barn offered a convenient seamark for navigating Herring Gut, a narrow channel near Bass Harbor, Maine. A cluster of "fish houses" identified Cape Cod's Race Point as three windmills, "which stand near each other upon an eminence," set Nantucket off from nearby Block Island and Martha's Vineyard (Furlong 1800:25, 54, 57, 64.). In the alongshore wilderness, mills, houses, and windmills acted as not only vital seamarks but outposts of civilization. These landmarks, however, were exceptions on the nation's eastern border. In fact, Zachary Lamson's aimless wandering on the featureless sand dunes of Cape Cod was a common experience for mariners wrecked on most sections of the American littoral (*American Citizen and General Advertiser* 1802; *Commercial Advertiser* 1802; *United States Oracle and Portsmouth Advertiser* 1802).

Frontier conditions characterized most of the early republic maritime cultural landscape.[9] Outside of port cities, coastal inhabitants lived in an isolated, parochial world secluded along the margins of a new nation, which was itself "at the very edges of Christendom, three thousand miles from the centers of Western

---

[8]The great exception on the desolate American coastline was the coast of Georgia and South Carolina, which was populated with sizable tidewater rice plantations (Steward 2002; Wood 2009:501–512, 527–528).

[9]Even the waterfront of Early Republic port cities were liminal zones, borderlands between maritime culture and a nascent American culture. It was a space inhabited by cultural mediators *par excellance*, sailors, and merchants (Gilje 2004).

civilization" (Wood 2011:274).[10] Poor roads and fitful waterborne connections physically isolated coastal communities. Resource extraction—farming, fishing, lumbering, and wrecking (salvage)—underwrote these communities' founding and continued existence. Even so, a growing number of settlements in the early republic littoral had moved beyond mere subsistence and were able to support a cadre of doctors, lawyers, craftspeople, and others who were not directly employed in resource extraction. Like other rural farmers, farmer-fishermen sought a modest competency and their participation in market exchange remained limited by the fact they consumed the bulk of what they caught and produced. Governmental and social constraints remained weak, dependent on the personal authority of prominent locals. Untamed wilderness—the sea—defined life in the American littoral. Indeed, the rise and fall of tides conditioned everyday life as seasonal weather cycles dictated the longer rhythms of coastal life throughout the nineteenth century (Henretta 1988; Kulikoff 1989; Vickers and Walsh 2005; Rockman 2009; McKenzie 2010).

Shipwrecks turned the American coast into a social and cultural frontier—a liminal space where different peoples and cultures converged, competed, and occasionally cooperated—because they routinely brought together three distinct groups: the urban owners and insurers of wrecked vessels; shipwrecked mariners, part of a cosmopolitan Atlantic maritime culture; and a fiercely local, isolated, and diverse frontier society.[11] *Industry*, for example, wrecked halfway between two villages. Lamson's experience with them illustrates the full spectrum—from familial-like cooperation to conflict—that social interactions could take in the liminal littoral. In Chatham, Lamson found "kind people" who "seemed to vie with each other to entertain me." In Orleans, he confronted rapacious locals who allegedly stole from the wreck (Lamson 1908:172–174). Lamson could not look to the federal government or a developed salvage industry for help—survival depended on individual initiative, the assistance of locals, and, maybe above all else, good luck. "Some etiquette between the [*Industry's*] underwriters and owners" left him without explicit directions for almost a month. Marine salvage remained an ad hoc affair, supervised by the wrecked captain or a knowledgeable local. Hired locals typically did the actual wrecking—removing the cargo, refloating the hull, or stripping the vessel of any valuable equipment. In fact, only after Lamson had finished salvaging the wreck did the schooner's owners hire a Cape Codder to oversee the salvage effort. All that was left for him to do was to pay the bills that Lamson accrued (Lamson 1908:171).

Industry wrecked during a transitional period in the history of the American maritime cultural landscape. Its shipwreck highlights the frontier reality of life in

---

[10]While the region produced many sailors who were among the most cosmopolitan Americans during this period, those who settled in the littoral lived in the insular world of their local community (Gilje 2004:26–32; Vickers and Walsh 2005; Howe 2007:40–41).

[11]This articulation of "frontier" is judiciously defined by Stephen Aron as "lands where separate polities converged and competed, and where distant cultures collided and occasionally coincided" (Aron 1994:128).

the sparsely populated, isolated, and parochial early republic littoral. Yet even as *Industry* wrecked, broader economic, political, and cultural processes were beginning to literally and figuratively knit the coastal frontier into the fabric of American society. Locals largely embraced this change because many were able to leverage their local knowledge and social connections to accrue financial and cultural capital in an increasingly connected coast. Maritime cultural landscapes like the early republic littoral are fundamentally dynamic and constantly evolving. As a theoretical construct, they offer one way to conceptualize the dynamic complex of space, place, patterns of habit, and thought of disparate historical actors. The rest of the chapter examines three interrelated factors driving this evolving American maritime cultural landscape: maritime commerce, enlightenment praxis, and the quest for a national identity. Shipwrecks, as we will see, were at the heart of every one of them.

Imperiled Revenue: Shipwrecks and the State

The significance of maritime commerce coupled with the dangers of the coast made shipwrecks and the littoral a central concern of the fledgling federal government, which depended on import duties as a critical source of revenue.[12] Too many vessels lost on the American coast undermined merchants' willingness to risk sending their vessels to American ports. Proactive legislators and federal officials used the tools of modern statecraft to begin to make the coast safer. The Customs Service, "the fiscal spine of the early United States (Gautham 2008:1)," began to systematically categorize and document—order—the sprawling, illegible American littoral for national policymakers. It also sent aggressive federal agents into the coastal frontier to protect the national interests threatened by coastal shipwrecks. Federally owned lighthouses and aids to navigation physically marked more of the coast every year, literally bringing light and order to a dark and dynamic frontier. Together these efforts mark the beginning of what would be a key factor in the transformation of the American maritime cultural landscape during the nineteenth century—federal intervention.

Nowhere did the early republic federal government act more decisively and explicitly to make a landscape legible than along the country's eastern frontier—its 1500 miles of coastline. The Customs Service, established in the summer of 1789, initiated federal efforts to regulate the Atlantic littoral. Data collection was at the core of the service's mission. As Secretary of Treasury Alexander Hamilton

---

[12]Between 1789 and 1800, customhouses collected $59.4 million, or almost 88% of $67.6 million in total federal revenue. Not until the Civil War would other revenue sources routinely match custom's receipts. Consequently, domestic politics, foreign relations, state and federal finances, as well as the international reputation of the new republic would be intractably tied to maritime commerce throughout the antebellum period (Wallis 2006; Gautham 2008:149–154; Wood 2009:93).

explained in his first circular letter to Collectors of Customs: "It is of the greatest moment, that the best information should be collected for the use of the Government" (Gautham 2008:101). Legislators divided the American coast into 59 customs districts—centered on the country's ports—run by appointed officials. The Collector of Customs, usually a prominent local figure, was responsible for keeping track of the shipping and cargoes landing in his district as well as the collection of duties. Depending on the size of his district, collectors could oversee a bevy of officers, including a "naval officer," inspectors, surveyors, weighers, and gaugers (Prince and Keller 1989:37–67). These appointed officials were "among the first and most numerous of federal employees" and their work created a prodigious number of documents, from Certificates of Registry and Manifests, to Crew Lists and Impost Books. These documents detailed the movement of goods and people along the coastal frontier, making the littoral (slowly, fitfully) legible to the new national government. Customhouses, however, remained bastions of localism rather than harbingers of federal authority when *Industry* wrecked (Labaree et al. 1998:168; Gautham 2008:95–100). Still, this important mechanism for making the littoral legible to national authorities existed and began to integrate an isolated frontier into the emerging national fold.

Coastal shipwrecks brought aggressive federal intervention in the coastal frontier. Wrecks containing foreign goods posed significant administrative and enforcement challenges for the Customs Service because they often occurred far from port city customhouses and the local context and connections on which the power of the Customs Service rested during the early republic. Instead of consisting of a mundane business transaction willingly undertaken by businessmen behind the closed doors of the customhouse, the exercise of federal authority became visible and coercive. Wrecking events damaged or destroyed dutiable cargoes, rendering meticulously drawn cargo manifests inaccurate and incomplete while casting potential revenue along miles of the coastal frontier. Customs inspectors dispatched to the scene made an accurate accounting of dutiable cargoes gathered from the beach or removed from a wreck. After describing every salvaged box, bale, hogshead, bundle, and article, inspectors appraised damaged goods and calculated appropriate duties. Inspectors worked closely with local wreckers and co-opted the power of local leaders, forging personal and institutional connections between the "outposts" of federal authority and the coastal frontier.

Every vessel entering every American port was a potential shipwreck that federal legislators wanted to safely guide through a customhouse.[13] State-owned and operated aids to navigation were the only feasible proactive measures then available to prevent shipwrecks. A week after establishing the Customs Service, Congress passed the "Light-House Bill," which effectively put the country's lighthouses and aids to navigation under federal control. The bill enabled individual states to cede

---

[13]Of course, customs officials were only interested in shipwrecks involving dutiable cargoes; revenue cutters did, however, assist distressed vessels regardless of the status of their cargo, although they were not explicitly ordered to do so until 1831 (U.S. Coast Guard 2011).

ownership of all existing "lighthouses, beacons, buoys and public piers" to the federal government. In exchange, the national government assumed all expenses related to their operation and maintenance. Unsurprisingly, this very Federalist bill garnered several days of heated discussion in the House of Representatives between states' rights advocates and Federalists led by Pennsylvania Congressman Thomas Fitzsimmons (*New-York Packet* 1789; U.S. Senate Historical Office 1991). The final version of the bill, a nuanced balance of sectional differences substantially influenced by Philadelphia merchants, took form in the Senate over 3 days of debate before becoming the seventh law passed by the first Congress. Coastal states, eager to decrease any financial and administrative costs, ceded every lighthouse and existing aid to navigation to federal control by 1795 (U.S. Congress 1789:51–56, 668–670; Weiss 1926:4–6; Putnam 1933:32).

The specter of shipwrecks made the coastal frontier a pressing concern for the country's highest officials. Dispatches and memoranda of successive presidents and cabinet members left a surprisingly robust paper trail concerning the minutia of lighthouse procurement, design, and everyday operation. George Washington "gave more than routine attention" to the administration of lighthouses, going so far as to inquire if part of an old mooring chain used to anchor a floating beacon in Delaware Bay could be saved and reused. Secretary of Treasury Alexander Hamilton personally directed many of the details of lighthouse work until delegating the task to the Commissioner of Revenue in 1792. That did not stop Thomas Jefferson from taking a personal interest in shipwrecks and lighthouse administration during his presidency. In a matter concerning the keeper of the Cape Henry Lighthouse in 1806, Jefferson wrote: "I think the keepers of light houses should be dismissed for small degrees of remissness, because of the calamities which even these produce; and that the opinion of Col. Newton in this case is sufficient authority for the removal of the present keeper" (Putnam 1933:33–38).

Congress willingly supported the nation's aids to navigation system, appropriating more than $550,000 for the matter during the Washington and Adams administrations. Easily passed, these early appropriations did not inspire the partisan politics that would make internal improvement bills the divisive matters they became later in the century because no one could dispute their vital importance to national survival (Larson 2001). The Light-House Bill was the first of 17 acts George Washington signed into law for the maintenance of existing aids, the construction of new lighthouses and beacons, and the deployment of distinctive red, white and black buoys at major ports, including Boston, New York, Portland, and New London. His successor, John Adams, was even more supportive, approving eleven measures totaling $234,052 during his 4-year term. Indeed, until President Jefferson approved a bill appropriating $30,000 for the construction of the Cumberland Road in 1806, all federally funded internal improvements were designed to make the American littoral legible to mariners. By 1800, 24 lighthouses —twice as many as 1790—guided mariners into the nation's principal harbors, bays, and sounds. Shipwrecks galvanized support for sustained federal investment in the American littoral that would profoundly alter the American maritime cultural landscape during the course of the nineteenth century (Risk 2011:24–47).

Despite such investment, federal presence in the turn-of-the-century littoral remained limited to a string of outposts on a desolate frontier. Coastal inhabitants retained a remarkable degree of autonomy through their knowledge monopoly and continued isolation. The decentralized administration of lighthouses, supervised at the local level by Collectors of Customs, left lighthouses (like customhouses) bastions of localism rather than harbingers of federal intervention when *Industry* wrecked in 1801 (Putnam 1933:39–40). Lighthouses, in other words, were federal buildings but local institutions. Locals similarly dominated coastal navigation. The 1789 Light-House Bill, for example, explicitly declined to regulate pilots, deferring instead to existing state statues. Navigation itself remained unregulated, subject to ancient, informal customs until the 1860s when Congress began to legislate the "rules of the road" (Palmer 2007:453–457).

Lighthouses and the Customs Service reflected Revolutionary leaders' broader efforts to put the "principles of the Enlightenment into practice" during the first decades of the American experiment (Gautham 2008:1; Wood 2009:4). Both were central to the federal government's initial efforts to centrally govern a socially, culturally, and geographically diverse republic by subverting localism and illegibility. The Customs Service began to order—make legible—the sprawling, prodigiously (arguably dangerously) dynamic early republic littoral as lighthouses literally brought light to the dark coast. Shipwrecks laden with dutiable goods brought federal agents into the coastal frontier where they forged durable relationships between coastal elites and the fledgling national government. Shipwrecks also spurred debate in the highest levels of government, bringing the first federal outposts—lighthouses—to the littoral and foreshadowing the significant federal presence that would define the American maritime cultural landscape during the second half of the nineteenth century.[14]

Enlightened Reform: Saving Lives, Codifying Knowledge

Shipwrecks inspired two significant nongovernmental efforts in the littoral that, like their federal counterparts, began to subvert the localism and illegibility of the early republic littoral. The erection of a series of lifesaving huts designed to succor the shipwrecked turned the most dangerous and isolated parts of the coastal frontier

---

[14]The federal government also focused on coastal defense beginning with the 1794 Naval Act. Coastal defense was "the historic twin of the Indian problem in American military history" according to Weighley (1984:98), and Congress voted to erect coastal fortifications and four arsenals for storing munitions. The War Department acquired sites through outright purchase or state succession while seaport residents contributed part of the necessary material and labor. Congress funded 764 rank-and-file to man the forts. By the War of 1812, 24 earthwork and masonry redoubts had been built along Atlantic Coast mounting 750 guns. These forts were located near seaports and did not expand the federal government's presence in the coastal frontier like customs inspectors and lighthouses did.

into the proper object of humanitarian reform. The publication of the first American-produced coast pilot, a specialized guide for mariners, marked the first sustained effort to codify and disseminate local knowledge of the coastal frontier. Like federal interventions along the coast, these private efforts were part of a sweeping post-Revolutionary reform effort rooted in Enlightenment ideals of order, humanitarianism, and the search for and dissemination of knowledge.[15] When *Industry* wrecked in 1801, these nascent efforts had done little more than establish a beachhead on a coastal frontier still dominated by entrenched locals, but they laid the foundation for efforts that collectively transformed the American maritime cultural landscape in the decades to come.

By the time *Industry* wrecked, a handful of crude wooden shelters built on the Massachusetts shore tangibly marked the coast with the fruits of post-Revolutionary enlightened reform. The first, built on Situate Beach in 1787, was a rough-hewn structure "erected from the principles of benevolence, to alleviate the distresses of the unfortunate shipwrecked seamen" by the Humane Society of the Commonwealth of Massachusetts (MHS).[16] Like later such shelters, it was stocked with candles, straw, fuel, and a tinderbox—basic survival items for mariners wrecked on Massachusetts's frigid, desolate shoreline. A 15-foot pole topped by a white ball identified each hut on the barren coast. Broadsides and newspapers announced their location and purpose to a wider audience, situating them on contemporaries' cognitive landscape of the maritime cultural landscape (*Boston Gazette* 1787; Humane Society of the Commonwealth of Massachusetts 1788). By the winter of 1801–1802, the MHS had built nine "humane houses" near the entrance to Boston Harbor, one in Plymouth, two on Nantucket, and one on the tip

---

[15]Enthusiasm for creating a better world seized post-Revolutionary America. While different sections of the new republic acted with differing opinions as to what that world would look like, many Americans saw their project as a vital contribution to the spread and fulfillment of the promises of the Enlightenment. They set about reforming and republicanizing their society, distancing it from its British origins and continuing the enlightened developments of the eighteenth century. As Wood (2009:470) writes, their goal was "to push back ignorance and barbarism and increase politeness and civilization." Private reformers worked to republicanize education and knowledge. They also formed charitable and humanitarian organizations and dramatically reformed the country's system of criminal punishment. Public initiatives dovetailed with these private efforts. Along the coastal frontier, state efforts to make the littoral more legible coincided with a range of enlightened private efforts to domesticate this illegible, premodern space (Gautham 2008; Wood 2009:4, 37, 469–507).

[16]The MHS was one of a cadre of humane societies in the turn-of-the-century Atlantic World motivated by Christian and Enlightenment values of universal beneficence. Founded in 1786 by Boston elites, the MHS was at the forefront of the "virtual explosion of philanthropic organizations" that began in the 1780s and continued into the nineteenth century (Wood 2009:470[quote], 470–495). Like other humane societies, the MHS combined the relatively broad membership of public charities, the international cooperation and exchange of knowledge of learned societies, and the moral responsibility to strangers advanced by groups from the Freemasons to abolitionists. It differed, however, in adapting the humane society movement's core mission (saving the drowned) to local conditions (few drownings; many shipwrecks) by constructing shelters for shipwrecked mariners and rewarding rescuers (Howe and Wolfe 1918:70–71; Heale 1968; Moniz 2009; Wood 2009:485–495).

of Cape Cod (Furlong 1800:39; Humane Society of the Commonwealth of Massachusetts 1846:1–10, 16; Howe and Wolfe 1918:57). The hut on Cape Cod, whose replacement Henry David Thoreau would later describe as a "seaside box," stood more than 20 miles from where *Industry* went ashore. It had been poorly constructed and "entirely demolished" by a storm the same month Zachary Lamson salvaged the remnants of his first command (Thoreau 1864:87–89).

Unlike the federal government, which delegated its coastal outposts to local control, MHS played a much more active role in the littoral. In the process, it forged nascent social and cultural bonds between the nation's urban core and coastal periphery. In 1802, for example, the MHS aggressively expanded its lifesaving network on Cape Cod by building six huts, including two that could have sheltered the crew of *Industry*. The location of these huts were based on a remarkable survey of the outer Cape conducted by the Reverend James Freeman, first minister of Boston King's Chapel and one of the founding members of the MHS. Freeman's 15-page report described the region's physical environment in intricate detail, synthesizing the collective knowledge of coastal locals and deep-sea mariners about the region's physical and shipwreck landscapes (Freeman 1802:17; Howe and Wolfe 1918:57).[17] Freeman analyzed the cape as a discrete maritime cultural landscape, identifying and codifying dangers for the MHS to mitigate through an analysis of the American littoral that state and federal officials would not undertake for several generations. In addition to surveying the Cape, Freeman forged agreements with eight Cape Coders—doctors, lawyers, ship captains, and reverends—to regularly inspect and report on the new huts, co-opting the local standing of prominent coastal residents to implement MHS's enlightened agenda. These were hardly imperialistic relationships—dozens of residents of Massachusetts's coastal frontier eventually joined the MHS, gaining cultural capital through the social bonds of humanitarian reform. As they did with federal incursions into their littoral, savvy locals embraced, shaped, and ultimately profited from outsiders efforts to alter and define the American maritime cultural landscape.

Freeman's report was exceptional—local knowledge of the American coast did not circulate widely at the beginning of the nineteenth century. Pilots, those with the most practical knowledge of the coastal frontier, held their knowledge of coastal waters dear not only to protect their profession but also because they were well aware of the impracticability of codifying knowledge of a constantly changing environment (Finger 2010). Other coastal residents with intimate knowledge of the littoral had few opportunities or incentives to disseminate their hard-earned knowledge. Few charts and coast pilots for the American coast were readily available when the *Industry* wrecked in 1801. According to one contemporary: "Here indeed few charts have been published, and those of no remarkable character either for the accuracy of their distances and bearing, or for the extent of their scale"

---

[17]Freeman's survey of the outer Cape resulted in half a dozen strategically located shelters that took into account topography, coastal erosion/sedimentation, weather patterns, coastal navigation, shipwreck patterns, and survivor psychology.

(Furlong 1800:vi). The most popular coast pilot, John Sellers's *The English Pilot, The Fourth Book* was similarly inaccurate, expensive, and it lacked the detailed local knowledge essential for safe navigation (Robinson 1969; Kreiger et al. 2001:94–96).

Coastal navigation—or gunkholing—drew on a rich body of accumulated knowledge and the mariner's senses. Mariners created an ever-evolving mental chart of the coastlines they sailed based on a lifetime of experience and handed-down knowledge. They literally felt their way along the shore; "following the shore line, just as in the earliest days of water transport, moving crabwise from rock to rock, from promontories to islands and from islands to promontories" (Braudel 1949:103). Tallow-dipped lead lines cast into the sea measured depth and sampled the composition of the seafloor. Changes in the color and temperature of the water, even the smell of pine trees and corn were elements in the sophisticated matrix mariners constructed to determine their position, assess their risk, and plot their course. This vernacular, practical, experiential knowledge, what Scott (1999:309–342) calls *metis*, was an eminently adaptable and successful way to navigate a dynamic physical environment. And yet, it was the very nature of this knowledge that made it so difficult to document and disseminate.

The publication of *The American Coast Pilot* in 1796 by Newburyport, Massachusetts, bookseller Edmund March Blunt marked the first sustained American effort to codify knowledge of the American littoral. Blunt intended the *Pilot* to be "a book of reference and direction" for mariners, and a relief "by the certain conviction that this cabin companion would be the means of security" for apprehensive owners of vessels plying the American coast.[18] Fear sells and Blunt frequently conjured the specter of coastal shipwrecks. "The life even of the most skillful and experienced mariner," he explained, "is more endangered as he approaches the coast … [than navigating] mid-ocean" (Furlong 1800:vi). *The American Coast Pilot* (at least theoretically) made the coast safer by collecting, for the first time, in "a compendious volume the most authentic descriptions of the harbours, and an accurate detail of the courses and soundings of the American coast." In this way, Blunt's search for and dissemination of knowledge was yet another Enlightenment-inspired effort that began to break down the isolation and localism of the coastal frontier at the beginning of the nineteenth century. In doing so he fundamentally altered the American maritime cultural landscape by bounding this dynamic, shape-shifting landscape between the covers of a book.

The American Coast Pilot provided the most detailed description of the early Republic's coastal frontier to date. Unlike *The English Pilot, The American Coast Pilot* was almost exclusively text and did not include any charts or sketches. But

---

[18]Blunt (1822:v–vi) later explained the intended and actual use of coast pilots was for vessels who were unable to get a local pilot, because the charts were not up to task and because printed, narrated directions were "vitally important" for safe navigation of coastal waters.

Blunt's *Pilot* offered a far more precise description of the many places American seafarers sailed: the Atlantic coast from Passamaquoddy to Key West, George's Bank, and the West Indies. Descriptions of harbors, bays, and well-used passages along the Atlantic coast were invaluable references for dead-reckoning mariners navigating unfamiliar waters. Not surprisingly, "remarkable" is frequently used in the text to describe many prominent features, places, and phenomena. Even more, the *Pilot* provides insight into the way mariners actually navigated. Natural (bushes, rocks, trees, hills) and human-made (lighthouses, buoys, and beacons, but also houses, barns, steeples, and windmills) features identified important waypoints. Mariners measured distance by the league (approximately three miles), the "biscuit throw," and the "musket shot." Vessels were simply "large" or "small," "loaded," "half-loaded," or "empty." And wind, tide, and time of day ultimately determined which anchorages were safe, which passages were prudent, and which coasts to avoid (Furlong 1800).

Blunt began updating the *Pilot* almost immediately after releasing the first edition, initiating a process of revision that occupied him for the next three decades. Revision involved forging and maintaining relationships with scores of mariners, customs officials, and pilots scattered along the coastal frontier. It also required Blunt to pay close attention to coastal shipwrecks, which, for him served as a bellwether for the presence of unknown rocks or shoals rather than imprudent decisions or perilous weather. The third edition of *The American Coast Pilot*, published in 1800, was a vital step in making the American littoral legible to outsiders because it combined multiple local knowledge into one printed reference for the first time. The 1796 edition had merely distilled the accumulated knowledge of New England mariners on whom Blunt relied for information. Its first entry, for example, "directions to sail into Boston," reflected the focus and knowledge of the *Pilot's* sources: ship captains primarily sailing in and out of Boston. The radically redesigned third edition tellingly begins with a description of Passamaquoddy—the northernmost point—and proceeds south mile by mile until concluding with sailing directions for the "Spanish Main." As Blunt boasted, the third edition was "a perfectly accurate compendium of the American coast navigation, combining all the information on this subject, which skillful experience and modern discovery have collected." It certainly appeared so. More than twice the size of the 1796 *Pilot*, the third edition corrected earlier errors, expanded descriptions of many parts of the coast, and included relevant state and local laws, as well as vital information about customs duties and procedures. Even so, it remained an option of last resort for the distressed and did not make mariners "their own pilots" as Blunt claimed it would— local knowledge simply remained "too intricate to describe" in print to outsiders (Furlong 1800:vi). Blunt would spend the rest of his life trying to capture in prose the ever-changing American maritime cultural landscape.

Shipwrecks, National Identity, and the Coastal Frontier

Despite the publication of *The American Coast Pilot*, the lifesaving huts of the MHS, and federal interventions along the coast, the American littoral remained spatially and culturally remote from the centers of national culture when *Industry* wrecked in 1801. Shipwreck narratives, however, brought the imagined littoral to the very heart of American life by helping Americans envision a national identity through the littoral. In the process, shipwreck narratives unwittingly began the process of culturally domesticating the country's eastern frontier—of capturing its harsh physical environment and mysterious population within the bounds of the new republic.

Turn-of-the-century America was awash in its maritime world and no maritime topic garnered greater interest than shipwrecks (Philbrick 1961; Stein 1975; Wharton 1979; Blum 2008; Cook 2013). Wrecks appeared in almost every issue of most early republic newspapers. As Lamson salvaged *Industry* in January and February 1802, news of 60 other wrecks on the American coast and dozens more on foreign shores appeared in American newspapers. Certainly, winter, with its intense storms, dangerous temperatures, and shorter days, was the peak season for maritime disasters. "Howling WINTER," wrote one perspicacious turn-of-the-century poet "when wrecks and beacons strew the steep,/and Specters walk along the deep" (Campbell 1802). Wrecks, however, did not strew the steep only in winter; mechanical and human error, severe weather, ignorance of currents and seafloor features, and simple bad luck were never precisely timed to the seasons.[19] Take the summer of 1801. A "tremendous hurricane" sent 120 vessels ashore near Nassau, New Providence in July (*Gazette of the United States* 1801). On the American shore: a brig wrecked in North Carolina at the end of April; seven drowned in the wreck of a Brooklyn ferryboat in May; Long Island claimed at least one vessel in June; a ship wrecked near Cape Henry in July; American warships rotted in the brackish water off Washington, D.C.; and a brig wrecked within sight of New York City (*Alexandria Advertiser and Commercial Intelligencer* 1801; *Federal Gazette and Baltimore Daily Advertiser* 1801; *The Bee* 1801; *The Courier* 1801; *The New York Gazette and General Advertiser* 1801a, b). The list could easily go on—ships, schooners, sloops, and brigs met disaster in the American littoral year round and newspapers reported every one about which they found out.

Most wrecks warranted only a terse statement in the "shipping news" column in the vein of one published by the *Salem Gazette* about *Industry*—"we are informed, that there were four vessels ashore on Cape Cod" (*The Salem Register* 1802). Occasionally, detailed letters, typically written by a shipwrecked mariner or passenger or sometimes by a local pilot to the vessel's owners, appeared in a separate column. These letters often provided graphic details about the fateful voyage and

---

[19]Every issue of *Lloyd's List* for the year 1802, for example, included reports of at least one (usually around a dozen) vessel wrecked, lost, condemned, damaged, on shore, or foundered at sea. This is impressive given *The List* was published twice a week, every Tuesday and Friday.

wreck, pandering to the growing influence of both sentimentality and a "predilection for scenarios of suffering" that permeated post-Revolution America (Halttunen 1998:68). More than sensationalist journalism or opportunities for schadenfreude however, these detailed letters disseminated news of the wreck to anxious relatives, merchants, insurers, and others interested in the fate of the vessels they described (Miskolcze 2007; Cook 2013). They also reflected a well-established American tradition—shipwrecks, along with any other news relating to ships, shipping, or the sea had been the "largest single news topic" for colonial papers (Copeland 1997:24).

Descriptions of shipwrecks in newspapers represent a fraction of the shipwreck narratives inundating the early republic. In fact, stirring narratives of loss and redemption set against the backdrop of harrowing storms, tempestuous seas, and lee shores were regular fodder for turn-of-the-century Americans. Chapbooks, magazines, pamphlets, broadsides, and plays recounting shipwrecks routinely appeared in American cities. Stories of maritime disasters thrilled audiences throughout the Atlantic world, and popular shipwreck narratives crisscrossed the sea. American booksellers advertised London editions of William Falconer's *The Shipwreck* and took subscriptions for American editions. *The Shipwreck: A Comic Opera*, first performed in London in December 1796, was a popular comedy routinely appended to longer five-act plays performed on stages between Boston and Charleston (*Claypoole's American Daily Advertiser* 1798; *City Gazette and Daily Advertiser* 1801). By 1806, nearly 200 book-length shipwreck narratives were available to an American public long accustomed to reading about the sea, including the first American edition of Archibald Duncan's seminal anthology, *The Mariner's Chronicle* (Philbrick 1961; Huntress 1974; Wharton 1979; Miskolcze 2007; Cook 2013).

While shipwrecks and the littoral existed on the margins of most people's everyday experiences during the early nineteenth century, they became a significant presence in their imaginations. Little surprise that citizens of a maritime state turned to shipwreck narratives to help them define and shape a national identity—itself a central concern of many post-Revolutionary Americans (Wood 2009:39–42; Smith-Rosenberg 2010; Wood 2011:274). Indeed, shipwreck narratives would become potent shapers and indicators of national identity for most of the nineteenth century because they dramatized and reinforced the notion of American exceptionalism, channeling national anxieties over what it was and where it was going (Philbrick 1961; Wharton 1979; Lane 2002; Misckolcze 2007; Cook 2013).

Shipwreck narratives also introduced and began to culturally domesticate—make legible—coastal people and places (Corbin 1995:1–19; Steinberg 2001:110–150). Newspaper accounts of coastal wrecks vividly described the harsh, desolate coastal environment; a place where sound vessels and competent seamen failed; a violent place dominated by wind and wave; a place represented as being beyond human control. And yet these accounts also worked to transform the coastal frontier into a landscape inscribed with identifiable places. Barnegat, Sandy Hook, Egg Harbor, and dozens of other places began to fill the nation's mental map of its eastern frontier. The wreck traps and safe havens described in shipwreck narratives became known kinds of places—semantically captured locales—on the margins of the new nation (Brückner 1999).

Early republic shipwreck narratives focused on the littoral's physical landscape but they remained mute on the people living on it. That newspaper accounts rarely mention "the inhabitants" of the American littoral attests to the lack of people living along the coast. But when they did, newspapers discussed coastal residents in generic terms and typically mentioned them because of the assistance they gave to the shipwrecked. A letter by the captain of the ship *Volusia*, wrecked 2 months after *Industry*, published in Boston's *Independent Chronicle* (1802) noted: "with some assistance from the inhabitants, we got on shore all safe." Less benevolent tales occasionally surfaced. For example, after the wreck of the ship *Ulysses*, immortalized by Michele Felice Cornè's 1802 painting *Capt. Cook Cast a Way on Cape Cod*, several newspapers reprinted the wrecked captain's "letter to his owner" in which he reported: "many of the inhabitants are bailing coffee and sand in their hats," i.e., plundering the wreck's cargo. The painting, however, showed the vessel as it wrecked on an empty beach (*The United States Oracle and Portsmouth Advertiser* 1802).

Works of fiction provided a more nuanced, if sensational, introduction to the inhabitants of coastal frontier. *The Shipwreck: A Comic Opera* is representative. A bawdy tale of shipwreck and plunder, the two-act comedy included nefarious wreckers, incompetent officials, helpless shipwreck survivors, and the lone benevolent local—the stock characters that would populate shipwreck narratives for the rest of the century (Arnold 1806). Like the myths and mythmaking of another frontier—the so-called "West"—myths about coastal inhabitants and places shaped the lived experience and historical development of the American maritime cultural landscape (Slatta 2010). At the beginning of the nineteenth century those myths were still being formed; not until the 1830s would the nefarious coastal wrecker become an American archetype. Fictionalized representations of coastal residents would play as significant a role in shaping public perceptions and state actions as the actual coastal populace would throughout the nineteenth century (Wells 2013). Their basic forms, however, were well known to Americans when *Industry* wrecked on Cape Cod.

Conclusion

The turn-of-the-century American littoral was an isolated, parochial, preindustrial space on the margins of the fledgling Republic. Yet the stage had been set for the wholesale integration of the American littoral with American life and thought. Indeed, the region's physical, social, and cultural landscapes were already being mapped by outsiders when *Industry* wrecked in 1801. The schooner's voyage, shipwreck, and salvage mark a transitional moment in the evolution of the American maritime cultural landscape as the coastal frontier foundered on the shores of enlightened reform, nation building, and sensationalist narrative.

Coastal shipwrecks like *Industry* are not only events, material remains, processes, or narratives that illuminate a maritime cultural landscape, but events, remains, processes, and narratives that defined and fundamentally shaped the American maritime cultural landscape. The vitality of maritime commerce and a cultural milieu that encouraged the enactment of Enlightenment ideals and articulation of a national identity turned twice-told tales of coastal shipwrecks into prime movers in the breakdown of the coastal frontier. At the highest levels of government, shipwrecks focused sustained attention on the coast because they imperiled the duties that financed government operations. In response, the federal government sent Customs officials into the coastal frontier and appropriated aids to navigation, the opening salvo in an aggressive and sustained policy of federal intervention in the littoral that continues to this day. Private initiatives mirrored these public efforts. By 1800, prominent, "enlightened" citizens in a handful of cities had begun to organize for the benefit of mariners wrecked on the nation's coastal frontier. At the same time, newspaper editors, artists, authors, and other cultural producers brought evermore tales of coastal shipwrecks into the homes and workplaces of Americans, introducing them to the wild coast and disaster through sensationalist narratives. Each of these groups mapped a part of the littoral, illuminating the American maritime cultural landscape as they subtly, irrevocably shaped it.

Still, the turn-of-the-century American littoral remained a frontier, linked by few roads, inhabited by few people, and rarely visited by outsiders. The federal government remained small and relatively toothless. Nascent state-level efforts to regulate the coast were just coming online when Zachary Lamson wrecked on Cape Cod. Americans were certainly becoming more familiar with their littoral, but it remained isolated and distinct from the center of national life—a distant frontier inhabited by rough frontier people on the fringes of law and society.

References

Alexandria Advertiser and Commercial Intelligencer. 1801. Shipping news. *Alexandria Advertiser and Commercial Intelligencer* 3 July:2. Alexandria, VA.

American Citizen and General Advertiser. 1802. Marine news. *American Citizen and General Advertiser* 27 February:3. New York, NY.

Arnold, Samuel James. 1806. The shipwreck, a comic opera. In *Cathorn's Minor British Theatre: Consisting of the Most Esteemed Farces and Operas, Volume the Sixth*, ed. John Cawthorn, 1–50. London: John Cawthorn.

Aron, Stephen. 1994. Lessons in conquests: towards a greater western history. *Pacific Historical Review* 63(2): 125–147.

The Bee 1801 New York. *The Bee 3 June: 2*. New London, CT.

Blum, Hester. 2008. *The View from the Masthead: Maritime Imagination and Antebellum American Sea Literature*. Chapel Hill, NC: University of North Carolina Press.

Blunt, Edmund. 1822. *The American Coast Pilot*, 10th ed. New York, NY: J. Seymour.

Boston Gazette. 1787. Mess'rs printers. In *Boston Gazette, 26 November: 2*. Boston, MA.

Braudel, Fernand. 1949. *The Mediterranean and the Mediterranean World in the Age of Philip II*. Reprinted 1995. Berkeley, CA: University of California Press.

Brückner, Martin. 1999. Lessons in geography: maps, spellers, and other grammars of nationalism in the early republic. *American Quarterly* 51: 311–343.

Burrows, Edwin G., and Mile Wallace. 1999. *Gotham: A History of New York City to 1898*. New York, NY: Oxford University Press.

Campbell, Thomas. 1802. Ode to winter. *The New Hampshire Gazette 9 February: 4*. Portsmouth, NH.

City Gazette and Daily Advertiser. 1801. Advertisement. In *City Gazette and Daily Advertiser 13 April: 3*. Charlestown, SC.

Claypoole's American Daily Advertiser. 1798. *Advertisement*. In *Claypoole's American Daily Advertiser 1 March: 2*. Philadelphia, PA.

Commercial Advertiser. 1802. Ship news. In *Commercial Advertiser 26 February: 3*. New York, NY.

Cook, Amy-Mitchell. 2013. *A Sea of Misadventures: Shipwreck and Survival in Early America*. Columbia, SC: South Carolina Press.

Copeland, David A. 1997. *Colonial American Newspapers: Character and Content*. Newark, DE: University of Delaware Press.

Corbin, Alain. 1995. *The Lure of the Shore: The Discovery of the Seaside, 1759-1840, Jocelyn Phelps, translator*. New York, NY: Penguin Books.

Dwight, Timothy. 1823. *Travels in New-England and New-York*, vol. 3. London, UK: William Baynes and Son.

Federal Gazette & Baltimore Daily Advertiser. 1801. In News. *Federal Gazette & Baltimore Daily Advertiser 18 June: 3*. Baltimore, MD.

Finger, Simon. 2010. A flag of defiance at the masthead: The delaware river pilots and the sinews of Philadelphia's Atlantic World in the eighteenth century. *Early American Studies* 8: 386–409.

Freeman, James. 1802. *A Description of the Eastern Coast of the Country of Barnstable*. Boston, MA: Sprague (Early American Imprints, Series 2, no. 2255).

Furlong, Lawrence. 1800. *The American Coast Pilot*, 3rd ed. Newburyport, MA: Blunt.

Gilje, Paul A. 2004. *Liberty on the Waterfront: American Maritime Culture in the Age of Revolution*. Philadelphia, PA: University of Pennsylvania Press.

Gautham, Rao. 2008. *The Creation of the American State: Customhouses, Law, and Commerce in the Age of Revolution*. Doctoral dissertation, Department of History. Chicago, IL: University of Chicago.

Gazette of the United States. 1801. News. In *Gazette of the United States 17 August: 3*. Philadelphia, PA.

Halttunen, Karen. 1998. *Murder Most Foul: The Killer and the American Gothic Imagination*. Cambridge, MA: Harvard University Press.

Heale, M.J. 1968. Humanitarianism in the early republic: the moral reformers of New York, 1776–1825. *Journal of American Studies* 2: 161–175.

Henretta, James A. 1988. Families and farms: *mentalite* in Pre-Industrial America. *William and Mary Quarterly* 35: 3–32.

Howe, Daniel Walker. 2007. *What Hath God Wrought: The Transformation of America, 1815-1848*. Oxford History of the United States: Oxford University Press, New York, NY.

Howe, Mark Anthony, and De Wolfe. 1918. *The Humane Society of the Common Wealth of Massachusetts: An Historical Review, 1785–1916*. Boston, MA: Riverside Press Cambridge.

Humane Society of the Commonwealth of Massachusetts. 1788. Advertisement. In *The Humane Society of the Commonwealth of Massachusetts*. Boston, MA (Early American Imprints, Series 1, no. 21156).

Humane Society of the Commonwealth of Massachusetts. 1846. *History of the Humane Society of Massachusetts*. Dickinson, Boston, MA: Samuel N.

Huntress, Keith. 1974. Introduction. In *Narratives of Shipwrecks and Disasters, 1586–1860*, ed. Keith Huntress, ix–xxxii. Iowa State University Press: Ames.

Independent Chronicle. 1802. Shipwreck. In *Independent Chronicle 8 March: 2*. Boston, MA.

Kreiger, Alex, David A. Cobb, and Amy Turner (ed.). 2001. *Mapping Boston*. Boston, MA: MIT Press.

Kulikoff, Allan. 1989. The transition to capitalism in rural America. *William and Mary Quarterly* 36: 120–144.

Labaree, Benjamin W., William M. Fowler, Jr., Edward Sloan, John B. Hattendorf, Jeffrey J. Safford, and Andrew W. German. 1998. *America and the Sea: A Maritime History*. Mystic Seaport Press: Mystic, CT.

Lamson, Zachary. 1908. *Autobiography of Capt. Zachary G. Lamson*. W.B. Clarke Company: Boston, MA.

Lane, Daniel W. 2002. *Nineteenth-Century American Shipwreck Narratives*. Doctoral dissertation, Department of English. University of Delaware: Newark, DE.

Larson, John Lauritz. 2001. *Internal Improvement: National Public Works and the Promise of Poular Government in the Early United States*. Chapel Hill, NC: University of North Carolina Press.

McKenzie, Matthew G. 2010. *Clearing the Coastline: The Nineteenth-Century Ecological and Cultural Transformations of Cape Cod*. Hanover, MA: University Press of New England.

Meinig, D.W. 1986. *The Shaping of America: A Geographical Perspective on 500 Years of History, Atlantic America, 1492–1800*. New Haven, CT: Yale University Press.

Miskolcze, Robin. 2007. *Women and Children First: Nineteenth-Century Sea Narratives and National Identity*. Lincoln, NE: University of Nebraska Press.

Moniz, Amanda Bowie. 2009. Saving the lives of strangers: humane societies and the cosmopolitan provision of charitable aid. *Journal of the Early Republic* 29: 607–640.

Mood, Fulmer. 1945. The concept of the frontier, 1871–1898: comments on a select list of source documents. *Agriculture History* 19: 24–30.

Mood, Fulmer. 1948. Notes on the history of the word *frontier*. *Agricultural History* 22: 78–83.

New York Gazette and General Advertiser. 1801a. Advertisement. In *New York Gazette and General Advertiser 25 July: 2*. New York, NY.

New York Gazette and General Advertiser. 1801b. Boston. In *New York Gazette and General Advertiser 15 August: 2*. New York, NY.

New-York Packet. 1789. Proceedings of congress. In *New-York Packet 18 July: 2*. New York, NY.

Oberb, Barbara B. and J. Jefferson Looney (editors). 2008. *The Papers of Thomas Jefferson Digital Edition*. University of Virginia Press: Charlottesville, VA. http://rotunda.upress.virginia.edu/founders/TSJN-01-27-02-0294. Accessed 18 May 2011.

Palmer, Sarah. 2007. Safety regulations for shipping. In *The Oxford Encyclopedia of Maritime History*, vol. 3, ed. John B. Hattendorf, 452–457. New York, NY: Oxford University Press.

Philadelphia Repository and Weekly Register. 1802. Lines on the winter of 1796. In *Philadelphia Repository and Weekly Register 27 February: 127*. Philadelphia, PA.

Philbrick, Thomas. 1961. *James Fenimore Cooper and the Development of American Sea Fiction*. Cambridge, MA: Harvard University Press.

Prince, Carl E. and Mollie Keller. 1989. *The U.S. Customs Service: A Bicentennial History*. Washington, DC: Department of Treasury.

Putnam, George R. 1933. *Lighthouses and Lightships of the United States*. New York, NY: Houghton Mifflin Company.

Risk, James. 2011. *Ship to Shore: Infrastructure and the Growth of American Seaports, 1790–1850*. Master's thesis, Department of History. College Park, MD: University of Maryland.

Robinson, A.H.W. 1969. Review of *The English Pilot: The Fourth Book* (London, 1689) by John Sellers. *The Geographical Journal* 135(2): 307.

Rockman, Seth. 2009. *Scraping By: Wage Labor, Slavery, and Survival in Early Baltimore*. Baltimore, MD: Johns Hopkins Press.

The Salem Register. 1802. Shipping News. In *The Salem Register 18 January: 3*. Salem, MA.

Scott, James C. 1999. *Seeing Like a State: How Certain Schemes to Improve the Human Condition Have Failed*. New Haven, CT: Yale University Press.

Slatta, Richard W. 2010. Making and unmaking myths of the American Frontier. *European Journal of American Culture* 29: 81–92.

Smith-Rosemberg, Carroll. 2010. *This Violent Empire: The Birth of an American National Identity*. Chapel Hill, NC: University of North Carolina Press.

Stein, Rober B. 1975. *Seascape and the American Imagination.* Potter, New York, NY: Clarkson N.

Steinberg, Philip E. 2001. *The Social Construction of the Ocean.* New York, NY: Cambridge Studies in International Relations. Cambridge University Press.

Steward, Mart A. 2002. *What Nature Suffers to Groe: Life, Labor, and Landscape on the Georgia Coast, 1680–1920.* Athens, GA: University of Georgia Press.

Stiles, T.J. 2011. *The First Tycoon: The Epic Life of Cornelius Vanderbilt.* New York, NY: Vintage.

Thoreau, Henry David 1864 *Cape Cod.* Reprinted 1984 by Parnassus Imprints, Hyannis, MA.

United States Oracle and Portsmouth Advertiser. 1802. Further particulars of the late dreadful shipwreck. In *United States Oracle and Portsmouth Advertiser 13 March: 3.* Portsmouth, NH.

U.S. Census Bureau. 1898. Population of the United States (Excluding Indians not Taxed): 1790 to 1820—11. In *Statistical Atlas of the United States, Based Upon the Results of the Eleventh Census* by Henry Gannett. Washington, DC: U.S. Dept. of Commerce.

U.S. Census Bureau. 2001. *Centers of Population for 1950, 1960, 1970, 1990 and 2000.* http://www.census.gov/geo/www/cenpop/calculate2k.pdf. Accessed 15 July 2012.

U.S. Congress. 1789. *Annals of Congress.* 1st Cong., 1st Sess.

U.S. Senate Historical Officers. 1991. *The Lighthouse Act of 1789.* Washington, D.C: U.S. Senate Historical Office. http://www.uscg.mil/history/docs/1789_LH_Act.pdf. Accessed August 15, 2012.

Vickers, Daniel, and Vince Walsh. 2005. *Young Men and the Sea: Yankee Seafarers in the Age of Sail.* New Haven, CT: Yale University Press.

Wallis, John Joseph. 2006. Federal Government Revenue, by Source: 1789–1939. Table Ea588–593. In *Historical Statistics of the United States, Earliest Times to the Present: Millennial Edition,* ed. Susan B Carter, Scott Sigmund Gartner, Michael R. Haines, Alan L. Olmstead, Richard Sutch, and Gavin Wright. New York, NY: Cambridge University Press. http://dx.doi.org.ezproxy.lib.uwf.edu/10.1017/ISBN-9780511132971.Ea584-678. Accessed January 1, 2016.

Weighly, Russel Frank. 1984. *History of the United States Army,* Enlg ed. Bloomington, IN: Indiana University Press.

Weiss, George. 1926. *The Service: Its History.* Activities and Organization: John Hopkins Press, Baltimore, MD.

Wells, Jamin. 2013. *The Shipwreck Shore: Marine Disasters and the Creation of the American Littoral.* Doctoral dissertation, Department of History. Newark, DE: University of Delaware.

Warton, Donald P. 1979. Introduction. In *In the Trough of the Sea: Selected American Sea-Deliverance Narratives, 1610-1766,* ed. Donald P. Wharton, 3–27. Westport, CT: Greenwood Press.

Wood, Gordon S. 2009. *Empire of Liberty: A History of the Early Republic, 1789-1815.* Oxford History of the United States: Oxford University Press, New York, NY.

Wood, Gordon S. 2011. The American Enlightenment. In *The Idea of America,* ed. Gordon S. Wood, 273–290. New York, NY: Penguin Press.

Writers' Project, Work Projects Administration, State of New Jersey. 1940. *Entertaining a Nation: The Career of Long Branch,* American Guide Series. Bayonne, NJ: Jersey Printing Company.

Chapter 4
Collaboration, Collision, and (Re)Conciliation: Indigenous Participation in Australia's Maritime Industry—A Case Study from Point Pearce/Burgiyana, South Australia

Madeline Fowler and Lester-Irabinna Rigney

Introduction

Understanding Indigenous peoples' formative role in early exploration and economic development of Australia throughout the contact and post-contact period is important. From first interactions with visiting mariners and shipwreck survivors, Indigenous peoples have been active agents within the maritime sphere. Evidence of Aboriginal maritime agency is also found in the archaeological literature of Aboriginal laboring in whaling, sealing, and pearling (McPhee 2001; Gibbs 2003; Paterson 2011). Another intersection between Aboriginal and maritime spheres occurred at coastal missions.[1] The missionary period began in Australia in 1823 with the establishment of missions in New South Wales (McNiven and Russell 2005:226). Despite their isolating agendas missions were, in many cases, still engaged with the maritime domain. Indigenous peoples living on missions across Australia built, owned, operated, and maintained boats (Roberts et al. 2013). The purpose of working vessels varied and included cultural obligations, transport, and

---

This chapter is dedicated to Narungga Elder Fred (Tonga) Graham (1932-2016) whose indomitable spirit, indefatigable good humour and commitment to his culture and country will be greatly missed. His honesty and integrity as a Narungga man never left him.

---

[1]The term 'mission' archaeology as used in Australia describes government reserves and institutions, including religious missions (Middleton 2010:182).

M. Fowler (✉)
Flinders University, GPO Box 2100, Adelaide, SA 5001, Australia
e-mail: maddy.fowler@flinders.edu.au

L.-I. Rigney
University of South Australia, GPO Box 2471, Adelaide, SA 5001, Australia
e-mail: Lester.Rigney@unisa.edu.au

© Springer International Publishing AG 2017
A. Caporaso (ed.), *Formation Processes of Maritime Archaeological Landscapes*,
When the Land Meets the Sea, DOI 10.1007/978-3-319-48787-8_4

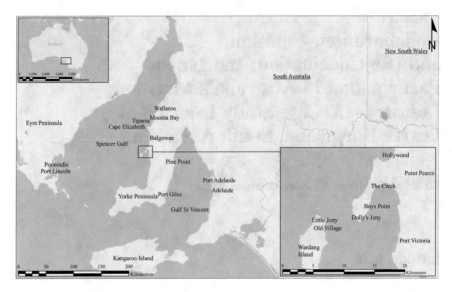

Fig. 4.1 Map showing Point Pearce/Burgiyana in relation to South Australia and Australia

fishing for subsistence and sale. Importantly, Aboriginal missions used colonial maritime networks for importing supplies, exporting products, transporting stock and people internally, as well as relying on marine resources for subsistence (Fowler 2013:74).

Inhibitors to Aboriginal maritime agency that harmed (or attempted to harm without always fully succeeding) Indigenous capacity for self-determination were the limitations and restrictions placed on the lives of Indigenous peoples. Regulation and restriction of human movement by the colonial administration, missionaries, and governments is evidenced in South Australia (Mattingley and Hampton 1992). These restrictions simultaneously acted on, and ignored, the maritime landscape. One mission that highlights this polarity is Point Pearce Mission/Burgiyana[2] located on Yorke Peninsula/Guuranda, the traditional land and sea of the Narungga people (Fig. 4.1).

The concept that Indigenous peoples played a minimal role in colonial maritime undertakings is firmly entrenched in traditional historiography. The objective of this chapter is to contribute toward the evolution of a broader Australian maritime archaeology by foregrounding the contribution of Indigenous peoples in Australia's colonial maritime industry (Roberts et al. 2014). Cross-cultural engagements will be explored through three main themes: in-kind transactions, maritime economy, and labor.

---

[2]The spelling of traditional Narungga words and names used in this chapter is consistent with the current orthography from the Narungga Aboriginal Progress Association (2006).

Literature Review

Westerdahl (2006:61) states that there was more rapid transfer of information at the coast and at ports and harbors (and, consequently, undoubtedly at coastal missions) than at inland areas. "Communication between the different coastal settlements have often been livelier than between the coastal settlement on one hand and inland settlements on the other" (Westerdahl 2003:20). The proximity of a busy international port, Port Victoria/Dharldiwarldu—recognized as a port in 1878—to Point Pearce/Burgiyana, established a decade earlier, played a significant role in the interaction of Narungga peoples with others and the speed at which the transfer of knowledge of Western maritime technology in the region occurred (Roberts et al. 2014).

Aboriginal and Torres Strait Islander Australians have made significant contributions to the Australian maritime industry and economy (Mullins 2012). The works of Wanganeen (1987) and Mattingley and Hampton (1992) illustrate that Aboriginal groups in coastal South Australia have a complex cultural, social, and economic relationship with the marine environment that both pre-dates British colonization and informs their engagement into colonial marine economies. These authors uphold the view that, similar to other coastal Indigenous peoples in Australia and elsewhere, Aboriginal groups like the Narungga in South Australia had well-defined coastal management plans tied to Dreaming stories[3] and communities of marine practices that sustained their terrestrial and marine resources. One important conclusion emerges from this analysis—that British colonization interrupts this robust cultural and economic relationship but does not extinguish it. There is no question settlement deals a major blow to these relationships; however, many Aboriginal groups like the Narungga are resilient and have continued to access and use traditional coastal and marine resources to the present day. These modern-day nuanced engagements within marine economies exist regardless of whether its purpose is for cultural affirmation or postcolonial economic survival.

Australian archaeological research under-represents Indigenous involvement in maritime activities (Roberts et al. 2013, 2014), although some archaeological research has been conducted in relation to Indigenous participation in sealing (Clarke 1996; James 2002; Russell 2005; Taylor 2008; Anderson 2016), whaling (Staniforth et al. 2001; Gibbs 2003), and pearling (McPhee 2001; O'Connor and Arrow 2008; Mullins 2012). Conversely, analyses of missions as a space for exploring maritime cultural heritage has largely been overlooked despite the role many mission individuals played in maritime, riverine, or lacustrine industries. While studies of missions have focused on spatial arrangement, material culture, fringe camps, and buildings (Birmingham 2000; Smith and Beck 2003; Ash et al. 2008; Jones 2009;

---

[3]According to Dixon et al. (2006:242), 'Dreaming' refers to 'a place or thing of special spiritual significance' and can include sites, stories, and paths or tracks. The 'Dreamtime', on the other hand, refers to the time of Ancestral Beings who created the world and environment and established moral codes (Dixon et al. 2006:241–242).

Dalley and Memmott 2010; Griffin 2010; Keating 2012), few studies have been undertaken on any maritime heritage centered at mission locations, with the exception of recent publications by the authors and their colleagues (Roberts et al. 2013, 2014; Fowler et al. 2014, 2015, 2016). Many missions in Australia are located on significant waterways, including coasts, rivers, and lakes; yet Indigenous interaction with these waterways at missions has not been fully researched archaeologically (Fowler 2013).

The preceding studies and this chapter suggest a need to posit an essential connection between Aboriginal and Torres Strait Islander peoples on coastal missions and colonial maritime history. Literature by other scholars concerned with this relationship through connecting non-Indigenous and Indigenous maritime activities in Australia include: visiting mariner (Mitchell 1996; Mulvaney and Kamminga 1999; Roberts 2004) and shipwreck survivor interaction (Morse 1988; Jeffery 2001; Nash 2006; McCarthy 2008; Merry 2010); rock art (Burningham 1994; Bigourdan and McCarthy 2007; O'Connor and Arrow 2008; May et al. 2009; Wesley et al. 2012; Taçon and May 2013); material culture and economy (Bowdler 1976; MacKnight 1986; Mitchell 1996; Gara 2013); and colonial labor forces (see aforementioned references to sealing, whaling, and pearling).

There is a need to contextualize cross-cultural engagements of Indigenous peoples with Western maritime culture. A maritime cultural landscape analysis is one means to bring this into focus. A widely accepted definition, which explicitly reveals the connection to maritime culture, is: "human utilization (economy) of maritime space by boat: settlement, fishing, hunting, shipping and, in historical times, its attendant subcultures such as pilotage, and lighthouse and seamark maintenance" (Westerdahl 1992:5). It is argued this Western approach targets colonial maritime activities in which Indigenous peoples were involved, and allows a new historiography—one that incorporates Indigenous worldviews—to be embedded.

The traditional maritime history of Australia reflects the hegemonic master narrative of the 'white' male that requires interrogation regarding its unexamined assumptions. The 'Aboriginal' question in maritime history requires quantification to reveal hierarchies of differences and the similarities between these two accounts of the past. As is the case with other countries (Westerdahl 2014:123), ships have been used as national icons in the national agenda, often reflecting great periods in the history of the ethnic identity of the colonizer (i.e., 'white') despite representing often brutal expansion policies. McGhee (1998) goes as far as to controversially suggest maritime archaeology paid homage to "ships symbolizing genocide." Maritime archaeology and its associated finds (ships) have been used by countries as metaphors for their great ages (e.g., 'Age of Discovery') (Westerdahl 2008:18). Attitudes of an amplified appreciation of Viking ships and a corresponding ignorance of Indigenous Russian (Slavonic) contexts, described by Westerdahl (2014:127), can be seen to be paralleled within Australian maritime heritage—at least that which is foregrounded in the domain of the general public. Discussions around traditional watercraft, such as the 2012 *Nawi* conference (Fletcher and Gapps 2012), are beginning to remedy a lack of archaeological research into Indigenous waterfaring, however it is the post-contact period that is significantly overlooked (Roberts et al. 2013:78). This attitude needs to be deconstructed; a maritime cultural landscape approach has been found to provide the useful tools to do so.

Brief Historical Background

The earliest interaction between Narungga and non-Indigenous peoples began with whalers and sealers in the 1830s, shortly followed by surveyors and pastoralists (Ball 1992:36; Mattingley and Hampton 1992:195; Krichauff 2008:51). Rigney (2002:xi–xii) describes such foreigner maritime 'collisions' using what he calls the ancient *First fire* narrative of the Gooreta/*Gurada* (Shark) Dreaming:

> The story talks of a meeting that was to be held by all the states of the Narungga Nation (Yorke Peninsula) at the head of Gulf St Vincent. A group of Narungga peoples caught a small fish and wrapped it in bark with its fins exposed. The fish was released in order that other groups around the Peninsula would catch the fish, indicating a grand meeting after the full moon of all Narungga. Several weeks later the fish returned swimming in and around a giant school. The children rushed into the water with screams of joy to catch the abundant fish. Soon mothers and fathers waded in knee deep to join the welcoming party. Screams of joy turned to screams of terror when the little fish was no longer small but huge. The fish had changed color from brown to white with razor sharp teeth. The small fish had turned into a large white pointer shark. The shark roared the Narungga word 'Bucha' (deadly/lethal). This story is a harrowing *First fire* prophecy that told our people to expect white danger from the sea. The genesis for this story could have been in response to any of a number of 'contact' observations and/or collisions, including sealers or subsequent expeditions by Colonel Light (aboard *Rapid*), Flinders (aboard *Investigator*), Baudin (aboard *Le Géographe*), or the crew of HMS *Buffalo*.[4]

Rigney (2002:xii) also suggests that the colonial maritime space is a controversial site of collaboration, collision, and reconciliation between Aboriginal and non-Aboriginal Australians. Yet most written narratives of this time echo all the assumptions about 'savages' and 'nature' that have dominated recent writing of Indigenous involvement in the colonial maritime industry. Even today this landscape of history is not one of a shared collective space either historically, philosophically, or intellectually. By contrast, investigating Aboriginal narratives provides mutual transformation of maritime archaeological studies and Aboriginal studies over and beyond the tensions that exist between them.

---

[4]Matthew Flinders left England in 1801 aboard *Investigator* to undertake a survey of the Australian coastline, including southern Australia. In 1802, Flinders encountered Nicolas Baudin, undertaking a similar expedition for the French, on board his ship, *Le Géographe*, at Encounter Bay in South Australia. *Rapid* left England for South Australia in 1836 with William Light, South Australia's surveyor-general, and his surveying expedition. South Australia's first governor, John Hindmarsh, also left England in 1836 as captain of *Buffalo*, which also transported other officials and settlers to the new colony (Gibbs 1969:11, 14, 30, 32).

Point Pearce Mission/Burgiyana was established in 1868, however during the late 1800s Narungga people were generally still mobile and were not restricted to the mission (Wanganeen 1987:25). In 1894, after the closure of Poonindie Mission, Indigenous peoples from the Eyre Peninsula joined Narungga residents (Wood and Westell 1998:8; Krichauff 2013:70). In 1915, following the 1913 Royal Commission into Aboriginal Affairs, it changed from operating as a mission by local trustees to being managed by the state government as a station. Aboriginal peoples' lives became increasingly regulated, although despite attempts, full control in this regard was not achieved (Kartinyeri 2002:70; Krichauff 2013:59). Point Pearce/Burgiyana people were involved in all aspects of station life including sheep shearing, farming, and building, and were also active on nearby Wardang Island/Waraldi in both pastoral activities and working for mining company Broken Hill Associated Smelters Pty Ltd (B.H.P.) (Wanganeen 1987:43, 55; Mattingley and Hampton 1992:118). The two main living areas on Wardang Island/Waraldi are the original mission outstation, known as the 'Old Village'—no longer in use as living quarters—and, to the north, the original B.H.P. township, still in use by the Point Pearce/Burgiyana community for accommodation. In 1966, the Point Pearce Aboriginal Reserve Land, including Wardang Island/Waraldi, became vested in the Aboriginal Lands Trust—ending government control—and has since been self-managed (Kartinyeri 2002:70).

Methods

Data collected for this research included oral history interviews with community members, archaeological surveys, and archival research. All data was collected under the guidance of Narungga people and in accordance with established ethical protocols.

Interviews were conducted with Narungga elders and community members using a semi-structured conversational style for initial interviews, to allow for flexibility in wording and question order (Fig. 4.2). Where feasible, participants were then taken to the sites under investigation. Therefore, the 'On Country' interviewing process became an experience for community members, whose responses to their interaction with the surrounding landscape elicited more textured and meaningful accounts of the past because of this in-landscape context. Place-based interviews were conducted at three main locations: the Old Village at Wardang Island/Waraldi, the Point Pearce/Burgiyana Peninsula coastline, and on-board a boat offshore from Wardang Island/Waraldi and Point Pearce/Burgiyana. See Fowler et al. (2015) for a more detailed discussion of the 'seeing land from the sea' in-landscape approach. 'On Country' interviews were also integral to identifying extant archaeological features.

Archaeological data collected include GPS positions and photographs of places, features, and artifacts; mapping of place names disclosed during the 'On Country' interviews; and producing site plans of jetties, a shipwreck, and the Old Village on

Fig. 4.2 Elder and fisherman Clem O'Loughlin with his boat at Point Pearce/Burgiyana. *Photo* by A. Roberts 11/25/13

Wardang Island/Waraldi. The focus of archaeological research on Wardang Island/Waraldi was to investigate the Old Village settlement and associated terrestrial and coastal maritime structures. Coastal sites/landscapes associated with maritime activity around the Point Pearce/Burgiyana Peninsula were surveyed with community members. Information recorded included place names, the activities that occurred at each location, and the people who were involved in those activities. An underwater snorkel and SCUBA survey was conducted at Boys Point/Gunganya warda. This location was identified in several of the oral history interviews as an area of intense and prolonged maritime activity, and therefore most likely for locating related material.

Archival research was conducted at several government repositories. The Trove database of the National Library of Australia was searched for historical newspapers relating to maritime activities at Point Pearce/Burgiyana. State Records of South Australia holds an extensive collection relating to Point Pearce/Burgiyana. This research effort focused on the earliest available unrestricted records (dating to 1909) related to maritime activities and the restricted series GRG52/1 'Aborigines' Office and successor agencies: Correspondence files,' the earliest relevant records of which date to 1881. The Port Victoria Maritime Museum provided material relating to Point Pearce/Burgiyana and Wardang Island/Waraldi, although the collections' general content are artifacts salvaged from the Wardang Island/Waraldi shipwrecks.

The National Museum of Australia's online digital collection was also reviewed. Secondary sources included local histories of Point Pearce/Burgiyana, many of which have Aboriginal authorship, the history of Yorke Peninsula/Guuranda written by non-Indigenous authors, and publically available genealogies.

Results

Westerdahl (2011:333) claims "it is not uncommon for multiple cultural landscapes [ethnic and economic cultures, social groups or communities] to exist in the same physical space," as is the case at Point Pearce/Burgiyana where Aboriginal people interacted daily with non-Indigenous missionaries, farmers, fishers, miners, and sailors. The Point Pearce/Burgiyana maritime cultural landscape reveals the shifting social and cultural structures, and technological changes, brought about by the contact and post-contact periods. In addition to subsistence and other cultural pursuits, maritime activities at Point Pearce/Burgiyana contributed to Australia's maritime industry through in-kind transactions, economy, and labor.

In-Kind Transactions

Traditional Ecological Knowledge of the Submerged and Coastal Environment

Indigenous peoples first contributed to the post-contact maritime economy through 'in-kind' support. It is suggested that initial contact between Narungga and non-Indigenous peoples, and the first encounter with boats, probably came with European sealers and whalers in the early 1830s, in the southwest of Yorke Peninsula/Guuranda (Mattingley and Hampton 1992:195; Wood and Westell 1998:3). The population of Indigenous women kidnapped by sealers from coastal South Australia and taken to Kangaroo Island may have included people from Yorke Peninsula/Guuranda (Russell 2005:2; Krichauff 2008:32; Taylor 2008).[5] Rigney (2013) also highlights that these visitors drew on Narungga knowledge of the maritime cultural landscape for navigation, which is often dismissed in the history of Kangaroo Island (Chittleborough et al. 2002; Rigney 2002).

The history of Narungga involvement in the navigation of Port Victoria/Dharldiwarldu is recognized by Narungga people as under-theorized and

---

[5]While Taylor's (2008:127–128) book does not feature any Narungga family names, women from the Walker family of Pt McLeay/Raukkan, who married into Point Pearce/Burgiyana families (Kartinyeri 2002:207), lived on Kangaroo Island, including Sally Walker.

Fig. 4.3 Little Jetty at Wardang Island/Waraldi. *Photo* by G. Lacsina 2/26/13

silenced (Rigney 2013). Aboriginal people were taken on-board European sailing ships to aid in piloting local areas and act as conduits to communicate with the local community. Indigenous people have been intentionally represented as just inter-preters for the local Aboriginal community, and their skills in navigation and seamanship are downplayed because it undermines the orthodox narrative of 'the great European sailor.' Aboriginal people assisted colonists in successfully navi-gating local maritime landscapes, although this traditional knowledge and under-standing is not credited in history.

Knowledge of local coastal topography, through cognitive or verbal seamarks, was one aspect of the Narungga community's expertise that was used by European colonizers. Such knowledge can only be accessed through place-based interviews as it is not recorded in the historical archives. A wide landscape of seamarks and transit lines was recorded during archaeological surveys at Point Pearce/Burgiyana and Wardang Island/Waraldi including natural seamarks, such as scrub and bushes, rocks and crevices, beaches, sand hills, mountains, and islands. Following colo-nization, cultural seamarks were incorporated into Indigenous knowledge systems, such as houses, telephone towers, fence lines, sheds, lighthouses, and jetties (Fig. 4.3). Roberts et al. (2013:84) document that an Aboriginal man named Ben Sims shared information about traditional fishing with a 'white' family of the same name (spelled Simms). The colonizers attempted to acquire sea knowledge, one of the currencies of power, whether such knowledge was imparted willingly or not. A more recent example of this, cited during oral history interviews, is non-Indigenous peoples attempting to 'take' fishing drops from Indigenous fish-erpeople by logging their marks while at sea.

Other traditional ecological knowledge adopted by Europeans included the location of freshwater. In 1878, drought conditions resulted in farmers and residents of Port Victoria/Dharldiwarldu traveling to the Point Pearce/Burgiyana Wells, in the vicinity of Hollywood—a coastal fringe camp—to source water (Moody 2012:17).

Trade of Traditional Resources for Western Goods

Indigenous peoples traded fish for needed (or desired) supplies with some of the first non-Indigenous peoples to travel into traditional Indigenous lands (Krichauff 2008:117, 124). Settlers, such as the artist and surveyor Edward Snell (at Yorke Peninsula/Guuranda in 1850), traded tobacco and red wafers[6] for Snapper, but-terfish, and leather jackets (Griffiths 1988:127–128). This was the beginning of Narungga people 'selling' (exchanging) fish within the Western economy. Snell also bought (exchanged) a pipe, tobacco, and a fourpenny knife for two Aboriginal fishing nets, most likely as curios rather than for their use (Griffiths 1988:128). In the first half of the twentieth century, Aboriginal man Joseph Edwards sold or traded fresh fish with local farmers (Ball 1992:38), "[He] used a net to catch fish, and on return he'd go around with a basket with a lid and a wet bag on it to cool the fish, and he'd go to all the farms with the fresh fish to sell" (Graham and Graham 1987:23).

Shipwreck Events at Point Pearce/Burgiyana

Point Pearce Mission/Burgiyana boats and Aboriginal residents were involved with shipwreck rescue events in the waters around Wardang Island/Waraldi. Following the wreck of *Aagot* (1907) on the west coast of the island, the stranded crew and their belongings were transferred to Port Victoria/Dharldiwarldu on the mainland in the Yorke Peninsula/Guuranda mission schooner, *Narrunga* (Advertiser 1907). Non-Indigenous Port Victoria/Dharldiwarldu pilot, Hector Orlando James Ximnez Simms, skippered the vessel and collected the crew from the landing at the northern end of the island, the crew having traveled across the island in a horse and dray owned by the mission (Moody 2012:81). While archaeological surveys to (re)locate the scuttled vessel *Narrunga* were unsuccessful, many tangible sites connected to its life history have been recorded archaeologically (Roberts et al. 2013). These intertidal and terrestrial sites include the Point Pearce/Burgiyana woolshed where the vessel was constructed, its launching location, Dolly's Jetty (Point Pearce/Burgiyana) and the Little Jetty (Wardang Island/Waraldi), and slipways on

---

[6]According to Krichauff (2008:117), these are probably slabs of dried pigment, as Snell was an artist.

Wardang Island/Waraldi. The slipways are an example of the everyday maintenance and upkeep of Point Pearce/Burgiyana vessels which are ignored in the
mission's archives. Personal anecdotes about such places recounted during 'On
Country' interviews aid in exploring a more textured account of these common-
place activities.

In February 1915, Leo Simms—a non-Indigenous man—was assisting with the
salvage of *Songvaar*, wrecked in the channel between Wardang Island/Waraldi and
the Point Pearce/Burgiyana Peninsula in 1912, when the boat he was traveling in
sank due to rough weather. He and the other crewman were rescued when two
16-year-old Aboriginal boys, "Stanley Smith and Clifford Edwards, saw the boat go
down and put to sea in a small dinghy from the Point Pearce jetty [Dolly's Jetty]
and rescued the two men" (Moody 2012:81, 261). While the Aboriginal connection
to Western vessels such as this are often not recognized, the Point Pearce School
donated a bell off the vessel *Songvaar*, presumably salvaged many years earlier, to
the Port Victoria Maritime Museum in 1972. Tangible objects such as this bell
reveal connections between the mission and the wider maritime industry in the area,
and in particular a continuing association with sites long after the event. Long-term
engagements are more readily understood through such curated objects than
through historical records, which only focus on major moments and events.

Maritime Economy

Fishing Economy

Commercial fishing at Point Pearce/Burgiyana, and smaller scale sale of fish to local
farms, has directly contributed to the development of Australia's fishing industry
and maritime economy. Indigenous peoples adapted resourcefully to colonial
pressures by integrating new technology, such as Western-style boats, and inter-
acting with the capitalist economy through selling products of their labor (Bennett
2007:86). Egloff and Wreck Bay Community (1990:28) note that most
non-Indigenous Australians do not realize the extent to which coastal Indigenous
peoples adopted European maritime technology. This is also recorded at the
Illawarra/Shoalhaven area of New South Wales where modified fishing by
Aboriginal people increased due to the introduction of fishing boats and nets by the
government, permitting a larger catch, and allowing surplus to be sold (Bennett
2003:257). As noted by Bennett (2003:260), it is not possible to quantify the
contribution the sale of fish made to Aboriginal subsistence because no records of
such transactions were kept.

Fishing is first recognized in the Point Pearce Mission/Burgiyana archives as a
form of employment for men in 1916 (Fig. 4.4) (South 1916:12). However, it was
undoubtedly a means of earning a living for themselves and their families prior to
this. Government assistance included being able to procure more suitable boats and
to install marine engines which allowed continuous employment in the commercial

Fig. 4.4 Clifford Edwards and Gilbert Williams in a Point Pearce/Burgiyana dinghy. Image courtesy of Jeffrey Newchurch on behalf of the Elaine Newchurch Collection

fishing industry (Aborigines' Protection Board 1948:2). In 1949, three fishermen received assistance to engage in commercial fishing, which brought the total number of Aboriginal people earning a living as fishermen at that time, and therefore independently from the station, to eight (Aborigines' Protection Board 1949:6). The last record of government assistance for fishing equipment dates to 1950 in the researched records, when the Station Manager stated:

> Two natives were provided with fishing nets to enable them to engage in fishing on a commercial basis. Other fishermen assisted in previous years are making a good living (Aborigines' Protection Board 1950:7).

The fringe camp Hollywood was a site of intensive fishing activities. Dinghy's used by people at Hollywood could be walked to shore in knee-deep water and four wooden poles, used for mooring boats, remain at this location (Fig. 4.5). Interestingly, oral histories record that both Aboriginal and other fishermen lived at Hollywood. The nature of Hollywood as a fringe camp means that the majority of references to it in archival sources regard it in a negative light and overlook the productivity and longevity of the maritime activities at this location.

Point Pearce/Burgiyana man Newchurch (2013) paints an evocative image of the daily routine of the fishermen at Boys Point/Gunganya warda, the primary moorings on the Point Pearce/Burgiyana coast:

Fig. 4.5 Four wooden poles at Hollywood used for mooring dinghy's, looking seaward (west). *Photo* by A. Berry 2/25/14

They would all go down there, gather at the Point, have their cigarettes and a yarn, their flagon of tea, cold tea, and their jam sandwich … plum jam, apricot … out they go, come in about five, six o'clock at night.

Terrestrial, intertidal, and underwater archaeological survey at Boys Point/Gunganya warda revealed evidence of this fishing industry in this area. A groove or cutting is visible in the bank which was used to push the boats in and out. Mooring posts, chain, and rope were also seen in the intertidal and submerged zone, with most mooring posts made of railway iron.

The underwater archaeological survey conducted at Boys Point/Gunganya warda also identified the remains of a fishing boat which sank at its moorings. Another Point Pearce/Burgiyana vessel, owned by Bart Sansbury, is held in the National Museum of Australia's National Historic Collection and its construction is very similar to the wrecked vessel at Boys Point/Gunganya warda (Hickson 2012:20), suggesting a similar type of fishing vessel was chosen by the Point Pearce/Burgiyana fishing community. Bart Sansbury's boat, however, is recognized for its historical rather than maritime significance (used in an exhibition to protest a lack of traditional fishing rights).

Newchurch (2013) also describes the attitudes toward fishing in his father's, Ronald Newchurch Senior, generation thus:

Fishing was their job, it was their employment. Fishing wasn't an escape from there [the mission]*; it was what they had to do to put food on the table.*

Commercial fishing and shore processing activities involved the whole family, including children, although the involvement of children and women is not recorded in historical archives. Net fishing occurred from boats and is recorded in several oral histories. Irene Agius remembers fishing around another mooring location, The Creek/Winggara, a freshwater creek that enters the sea north of Boys Point/Gunganya warda:

> We used to just stand up on top of the dinghy and pole along … And they'd throw bread out to feed the garfish, all stale bread … Net fishing. And we'd have one of the big old tubs, you know what you have a bath in? That was brought down in a jinker … and we'd come along in the buggy, all the kids. And then night, when he'd (her father) bring the fish in, we'd be there, stripping them you know. All the garfish. And putting the little ones one side and the bigger ones one side. Oh yeah, tedious job. … We'd just take the whole lot, put 'em all back in the bag and he'd take them all down to the Port [Victoria] and sell them. Used to get … £5 a bag. Wheat bag, you know, … So he'd get three bags, four bags … We'd have our feed on the coals here … Never worry about … scaling them …, cleaning them. You'd rinse them out and stick 'em on the coals and then just eat the flesh. … [we ate] fish nearly every day. … If fish was there we'd eat it (Wood and Westell 1998:14–15).

This memory is similar to those recorded during oral history interviews where community members remembered waiting at The Creek/Winggara as children for fathers and grandfathers to come in at night. Children also fished from boats with family members, undertaking tasks, such as pulling up the anchor, cleaning the boat down, and stripping, cleaning, and packing the fish ready for sale to the wholesaler at Port Victoria/Dharldiwarldu.

Rigney (2013) suggests that more and more Narungga people bought Western boats because they needed to go further out due to demand on fish being so high in the traditional fishing places as a consequence of the growth in the Western commercial fishing market. More Narungga people had to go off mission, to areas such as Balgowan, Cape Elizabeth, Tiparra/Dhibara, Moonta/Munda Bay, and around Yorke Peninsula/Guuranda, to compete with recreational fishermen (Rigney 2013). Small-scale fishing enterprises, such as those run at Point Pearce/Burgiyana, were impacted heavily when nonlocal fishermen began using larger nets in local waters. Traditional fishing at Point Pearce/Burgiyana is being further explored by Roberts et al. (in prep).

Buying and Selling Boats and Maritime Equipment

Indigenous people also participated frequently in the local economy of purchasing and selling boats and other maritime equipment such as sails and nets. As noted by Rigney (2013), the modern interpretation of understanding boats came when Aboriginal peoples began to buy, and build (Roberts et al. 2013), Western-style, non-Indigenous boats. Assistance in relation to fishing activities occurred across South Australia, not just at Point Pearce/Burgiyana. The 1915 Chief Protector of Aboriginal's report, states that many Aboriginal people "have been assisted in the purchase of boats, guns, fishing nets, seed, wheat etc. with but little good resulting"

(South 1915:3), dismissing the participation of Aboriginal people in a healthy industry for equipment manufacturers, distribution agencies, and markets.

Aboriginal people at Point Pearce/Burgiyana began owning their own fishing boats and dinghy's from at least as early as 1895, as suggested by the file title which describes a request of payment for the freight of John Milera's boat to Moonta/Munda from Port Lincoln on a steamer (GRG52/1/209/95).[7] This was following his relocation from Poonindie Mission to Point Pearce Mission/Burgiyana (Kartinyeri 2002:374). Aboriginal people were seldom able to apply for loans to purchase boats on their own, usually being required to have some form of recommendation by a non-Indigenous member of society. Two letters highlight this, the Point Pearce/Burgiyana superintendent's recommendation for the supply of sail for Fred Graham costing £3.0.0 (GRG52/1/38/15) and a resident of Port Victoria/Dharldiwarldu recommending a boat for Philip Welsh of Point Pearce/Burgiyana (GRG52/1/298/07). Boats were generally purchased from non-Indigenous fishermen in nearby towns, such as the letter in which Robert Wanganeen applied for assistance to purchase a boat at Moonta/Munda (GRG52/1/41/96).

Other materials for vessels, such as anchors, oars, mooring chain, and motors went through the same process as requests for boats:

> Alfred Hughes—Point Pierce (Pearce)—Asking for 1 12lb anchor and a pair of oars for his boat (GRG52/1/64/14).

> Joseph Yates—Aboriginal—Point Pierce (Pearce)—Applies for and encloses 20/- in past payment for chain for mooring his boat (GRG52/1/408/98).

> Harold Kropinyeri—Half Caste Aboriginal—Loan of £35 to enable him to purchase a motor for his boat. Cost £65 (GRG52/1/Jan-13).

The purchase of these goods came from major shipping centers, such as Port Adelaide, as attested to by the letter from Charles Adams requesting to be supplied with a fishing net at the cost £8.12.0 from Port Adelaide plus 5% and transport, totaling £9.10.0 (GRG52/1/56/14), and further contributed to the demand for intrastate transport. In all of these cases, it can be seen that the cost was considered to be highly important, given that it is often listed in the files' titles, highlighting the capitalist system within which Aboriginal people were required to participate.

Supplying Goods for Coastal Trade

Point Pearce/Burgiyana is part of a wider network of trade and economics. Goods from the mission were also supplied to, or transported via, the coastal sea trade. Pig

---

[7]The Correspondence files of the Aborigines' Office and successor agencies (GRG52/1) are restricted because they contain sensitive material. Only the title of files, which are available to the public at State Records of South Australia, are reproduced here as the contents of the files are not allowed to be published.

breeding activities at Point Pearce/Burgiyana directly contributed to the international sailing economy by acting as a supplier of meat to the vessels. In 1933, the wheat ships loading at Port Victoria/Dharldiwarldu purchased a number of pigs from Point Pearce/Burgiyana via the local butcher (McLean 1933:9). Presumably, these were for the consumption of the ships' crew.

The production of wool and wheat at Point Pearce/Burgiyana, and the Indigenous labor in both of these industries, all silent or underplayed in Western literature, indirectly contributed to Australia's coastal trading economy and by extension the international distribution of Australian goods. This includes the transport of mission produce, including wool and wheat, via coastal traders to major centers. The wool from the 1918 shearing on Wardang Island/Waraldi was shipped to Port Adelaide on the ketch *Alert* (GRG52/73/1). In addition to archaeological survey of the two aforementioned jetties, survey also targeted the 'Old Village' on Wardang Island/Waraldi, which is where the Point Pearce/Burgiyana families who cared for the sheep that were grazed and shorn on Wardang Island/Waraldi lived (Roberts et al. 2013; Fowler et al. 2014). Archaeological evidence in this area is related to both daily domestic life, such as living quarters, ceramic, glass, brick, and bone artifact scatters, and the pastoral activities including infrastructure such as in-ground and above-ground tanks for the water supply (Fowler et al. 2014). The maritime cultural landscape of Point Pearce/Burgiyana is deeply involved in other occupations such as coastal and island agriculture and pastoralism. Therefore, the investigation of cultural activities relating to pastoralism contextualized the maritime focus, where the same people at Point Pearce/Burgiyana were involved in both maritime and pastoral cultures on the coast—fishing farmers or farming fisherpeople.

In another instance, shearing from 1932 was dispatched per the coastal ketch *Lurline* (1873–1946) (GRG52/70). It appears that the grain and wool from Point Pearce Station/Burgiyana was lumped to Port Victoria/Dharldiwarldu for transport by ship; however, there are references to goods being transported direct from Dolly's Jetty, which saved on the long cartage through to Port Victoria/Dharldiwarldu. The letter from the superintendent of Point Pearce/Burgiyana asking for arrangements to be made by the Wheat Harvest Board for 9000 bags of wheat to be shipped from Dolly's Jetty (GRG52/1/92/1916) gives some indication of this.

Labor

Working at Ports

Lastly, Indigenous peoples have provided labor to Australia's maritime industry. As discussed, Indigenous participation in the maritime industry at Port Victoria/Dharldiwarldu has been ignored and silenced in Western histories (Roberts et al. 2014:29). However, in addition to going out to different communities for

Fig. 4.6 "At Point Pearce where the men lumped bags of wheat onto boats for shipping." Image courtesy of the Dr. Doreen Kartinyeri Collection, South Australia Native Title Services

seasonal work, there were Indigenous peoples that worked in and around Port Victoria/Dharldiwarldu, and at other ports on Yorke Peninsula/Guuranda, in relation to shipping. Evidence of this has been found through oral histories and, to a more limited degree, archival research. Lumping—the occupation of carrying heavy loads onto ships—at a number of ports around Yorke Peninsula/Guuranda is a key area (Fig. 4.6).

Narungga Elder Clem O'Loughlin's brother, Alfred 'Locky' Jnr, lumped for *Passat*, *Lawhill*, and *Pamir* in the 1940s, taking bags on and off the jetty to the boats (Roberts et al. 2014:28–29). In addition to overseas ships such as these, lumping of wheat at Port Victoria/Dharldiwarldu also included coastal ketches, schooners, and steamships bound for Adelaide, such as SS *Nelcebee* in the 1940s and 1950s. Narungga people worked at ports at other places on Yorke Peninsula/Guuranda other than stacking and lumping at Port Victoria/Dharldiwarldu.

In 1916, all the wheat lumping at Balgowan Jetty, north of Point Pearce/Burgiyana, was done by Point Pearce/Burgiyana men and again in 1919, J.B. Steer, the Superintendent, recorded most lumping was done by Point Pearce/Burgiyana men, although very little wheat was shipped from the port that season (South 1916:12, 1919:11). The port of Balgowan frequently gave work to men from Point Pearce/Burgiyana as wheat lumpers, including again in 1920 and 1921 (Garnett 1920:10, 1921:10). Other ports on Yorke Peninsula/Guuranda where Point Pearce/Burgiyana people are recorded as working, through oral history interviews, include Pine Point and Port Giles on the eastern coast and Wallaroo/Wadla waru on the western coast. Point Pearce/Burgiyana people also lumped as far afield as the major South Australian port of Port Adelaide, such as for the coastal vessels *Annie Watt* and *Falie*. Fred Warrior for example, who grew up at Point Pearce/Burgiyana, did "wharf work" in the 1960s (Warrior et al. 2005:113).

Working as Skippers

Indigenous peoples worked as skippers on several boats which were used for other enterprises, which have occurred on Wardang Island/Waraldi over the past century, including the B.H.P. launches. For example, *Silver Cloud* (built in 1942 as a flying boat tender), a motor launch just under 40 ft. (12.19 m) in length, was originally used by B.H.P. and skippered by B.H.P. employee Jack Doyle (Heinrich 1976:110). Used to transport tourists and cargo between Wardang Island/Waraldi and Port Victoria/Dharldiwarldu, Aboriginal men Clem Graham and 'Nugget' Rankine also skippered and operated it (Sunday Mail 1974). Clem Graham was one of the first two Aboriginal people in South Australia to get a skippers certificate (Anonymous 1960).

Discussion

Reynolds (1984) notes that the "Great Australian Silence"—"a cult of forgetfulness practiced on a national scale" (Stanner 1969)—was a twentieth century phenomenon which has been broken since 1968 (following the 1967 Commonwealth referendum). It is argued, however, that the Indigenous experience in the maritime industry is still "slowly, unevenly, [and] often with difficulty" being incorporated into the image held by non-Indigenous Australians of their national past (Reynolds 1984). The oral history, archaeological, and archival results presented challenge the (still popular) view that pioneering was the exclusive achievement of Europeans and that Aboriginal peoples contributed nothing to the colonization of the country (after Reynolds 2000:287).

Oral history is instrumental in recording the archaeological landscapes of traditional knowledge, including seamarks. Some constructed places at Point Pearce/Burgiyana, such as buildings and fence lines, were not known to be part of the maritime cultural landscape until their identification by community members as seamarks. Furthermore, a lack of purpose-built seamarks has resulted in an almost exclusive use of verbal seamarks, only documented through community-based, collaborative research. It is this cognitive knowledge which became part of transactions with European colonizers.

The assistance of Aboriginal people during shipwrecking is also documented at Point Pearce/Burgiyana, as well as a continued relationship to such wrecks through salvage activities. These Aboriginal connections must be recognized when celebrating anniversaries of these wrecking events (e.g., 100th anniversary of the wrecking of *Songvaar* in 2012). Similar celebrations in recent years, such as the 2002 bicentenary of the meeting of Flinders and Baudin in South Australian waters, have notably lacked Aboriginal participation or an interrogation of the Aboriginal presence in such events (Rigney 2002:xii–xiii).

Ships in the past, in different places of the world, have been employed as symbols of power (Westerdahl 2008:25). The mission's control at Point Pearce/Burgiyana extended significantly to the purchase of boats by individuals, ensuring power in this domain; for example, the simple necessity that Aboriginal people had to "request," "apply for assistance," and "ask for" loans to purchase vessels and other equipment. By encouraging dependency on wage labor and Western goods through stimulating the maritime industry—including investing money in jetties and providing loans for boats and fishing equipment—the colonial missions and government attempted to manipulate the Indigenous maritime landscape to promote a capitalist worldview (Meide 2013:16, 22).

While there appears to be control in regard to acquiring fishing vessels, it seems that missionaries/government did not attempt to control the fishing activities themselves, in terms of claiming the catches, in comparison to agriculture, such as the share farming system. The sale of fish was made directly with the wholesaler at Port Victoria/Dharldiwarldu with Aboriginal people receiving the entire profit, although this may not have been equal to that earned by non-Indigenous fishermen. Working directly for the mission, Aboriginal people received a share of the wheat and wool profits, which was not equal to what they would have been entitled, had they been farming independently.

The professional fishing activities at Point Pearce/Burgiyana were, for the most part, a shore-based activity; however, this world was very much entangled with the everyday and non-professional, i.e., cultural fishing activities. Oral histories reveal that children were involved in the launch of workboats, working as 'deck hands,' and cleaning fish in preparation for sale. Thus, children were also gaining invaluable practical experience and cultural knowledge at the same time as being 'on the job' [see Roberts et al. (2014) for more discussion on children in the maritime domain at missions].

The final focus of this discussion, Aboriginal labor, enables the identification of wider maritime cultural landscape transport zones and sea routes, and allows the research to engage with the racialized aspects of the maritime industry (Croucher and Weiss 2011:8). Furthermore, the location of the wrecked vessel at Boys Point/Gunganya warda reinforces both the tangibility of sea routes and the highly dynamic nature of landing sites as centers of maritime activity. It is therefore imperative to envisage wider landscapes, including networks of trade and economics, and partner economies to the maritime industry (e.g. agriculture, pastoralism, and mining), when considering the maritime cultural landscape approach for Indigenous post-contact contexts. By doing so, a more complex story of engagement emerges between Indigenous and non-Indigenous peoples who do not readily occupy predetermined niches in the landscape, for example highly mobile Aboriginal fisher-farmers (after Gill et al. 2005:126). Finally, this research has not only foregrounded Indigenous contributions to the maritime industry but has also been able to put names to those Aboriginal individuals who were involved, through both oral histories and archival documents.

Conclusions

This chapter's research is a collaborative authorship between a Narungga and non-Narungga scholar. It has sought to understand how a maritime cultural landscape approach contributes to reinstating Indigenous visibility in Australia's maritime heritage. The authors argue that the findings aspire to transform the discipline of Australian maritime archaeology by foregrounding the contributions of Indigenous peoples in Australia's colonial maritime industry. The maritime cultural landscape provides an alternative historical and geographical framework which, in this instance, has formed the basis for a more inclusive local, regional, and national story (Gill et al. 2005:127).

The research has provided an unprecedented insight into the world of Aboriginal peoples in the post-contact maritime landscape and has allowed for the presence of Narungga people in a heretofore Western maritime literature. Maritime and Indigenous archaeology do not occupy separate physical spaces, in many cases they overlap, which poses a challenge to maritime archaeologists to comprehensively record both spheres. The archaeological evidence of the maritime cultural landscape of Point Pearce/Burgiyana reveals the tangible connections to traditional knowledge, the everyday and unremarkable aspects of daily life, and the longevity of maritime activities over time. Such aspects are overlooked in the capitalist, event-focused, and controlling purview of the mission's archives, however are also deeply enriched through narratives recounted by community members.

The maritime cultural landscape framework, employed here, could be used to investigate other themes of cultural contact, such as Macassan trepang processing stations, shipwreck survivor camps, shore-based whaling, sealing and fishing stations, major pearling ports, and shipbuilding centers, to continue to foreground Indigenous peoples participation in the Australian maritime industry. Comparative studies could also occur at missions situated on other waterways (for example rivers or lakes) and in similar contexts in other countries (for example countries in the Pacific region). The maritime cultural landscape framework has offered some small part in providing an antidote to colonizing landscapes (Gill et al. 2005:3). It has facilitated the telling of a more complicated story (Reynolds 2000:9).

Acknowledgments The authors are grateful to Amy Roberts, Klynton Wanganeen, and the editor of this volume for their time in providing constructive comments during the drafting phase of this chapter. The authors would also like to thank the Narungga Aboriginal Corporation Regional Authority, Narungga Nation Aboriginal Corporation, Point Pearce Aboriginal Corporation, and Adjahdura Narungga Heritage Group for their involvement and support of this research. Also thanks to Point Pearce/Burgiyana elders and community members who participated in this project as heritage monitors and shared their knowledge during interviews. This research would not have been possible without the grant received from the Professor Ronald M. and Dr. Catherine H. Berndt Research Foundation (2013) and the scholarship received from the Australasian Institute for Maritime Archaeology (2013). This research was approved by the Flinders University Social and Behavioural Research Ethics Committee (Project 5806).

References

Aborigines' Protection Board. 1948. Report of the Aborigines' Protection Board for the Year Ended 30th June, 1948. K.M. Stevenson, Adelaide, SA.

Aborigines' Protection Board. 1949. Report of the Aborigines' Protection Board for the Year Ended 30th June, 1949. House of Assembly, Adelaide, SA.

Aborigines' Protection Board. 1950. Report of the Aborigines' Protection Board for the Year Ended 30th June, 1950. K.M. Stevenson, Adelaide, SA.

Advertiser. 1907. Wardang Island Wreck: Vessel Breaking Up. *Advertiser* 15 October:10. Adelaide, SA.

Anderson, Ross. 2016. Beneath the Colonial Gaze: Modelling Maritime Society and Cross Cultural Contact on Australia's Southern Ocean Frontier—the Archipelago of the Recherche, Western Australia. Doctor of Philosophy thesis, Department of Archaeology, University of Western Australia, Crawley, WA.

Anonymous. 1960. Aborigines are Skippers.

Ash, Jeremy, Alasdair Brooks, Bruno David, and Ian J. McNiven. 2008. European Manufactured Objects from the 'Early Mission' Site at Totalai, Mua (Western Torres Strait). *Memoirs of the Queensland Museum Cultural Heritage Series* 4(2): 473–492.

Ball, Megan. 1992. The Lesser of Two Evils: A Comparison of Government and Mission Policy at Raukkan and Point Pearce, 1890–1940. *Cabbages and Kings: Selected Essays in History and Australian Studies* 20: 36–45.

Bennett, Michael. 2003. For a Labourer Worthy of His Hire: Aboriginal Economic Responses to Colonisation in the Illawarra and Shoalhaven, 1770–1900. Doctor of Philosophy thesis, University of Canberra, Bruce, ACT.

Bennett, Michael. 2007. The Economics of Fishing: Sustainable Living in Colonial New South Wales. *Aboriginal History* 31: 85–102.

Bigourdan, Nicolas, and Michael McCarthy. 2007. Aboriginal Watercraft Depictions in Western Australia: On Land, and Underwater? *Bulletin of the Australasian Institute for Maritime Archaeology* 31: 1–10.

Birmingham, Judy. 2000. Resistance, Creolization or Optimal Foraging at Killalpaninna Mission, South Australia. In *The Archaeology of Difference: Negotiating Cross-Cultural Engagements in Oceania*, ed. Robin Torrence, and Anne Clarke, 361–405. London, UK: Routledge.

Bowdler, Sandra. 1976. Hook, Line and Dilly Bag: An Interpretation of an Australian Coastal Shell Midden. *Mankind* 10(4): 248–258.

Burningham, Nick. 1994. Aboriginal Nautical Art: A Record of the Macassans and the Pearling Industry in Northern Australia. *The Great Circle* 16(2): 139–151.

Chittleborough, Anne, Gillian Dooley, Brenda Glover, and Rick Hosking (eds.). 2002. *Alas, for the Pelicans! Flinders, Baudin and Beyond*. Kent Town, SA: Wakefield Press.

Clarke, Philip A. 1996. Early European Interaction with Aboriginal Hunters and Gatherers on Kangaroo Island, South Australia. *Aboriginal History* 20: 51–81.

Croucher, Sarah K., and Lindsay Weiss. 2011. The Archaeology of Capitalism in Colonial Contexts, An Introduction: Provincializing Historical Archaeology. In *The Archaeology of Capitalism in Colonial Contexts: Postcolonial Historical Archaeologies*, ed. Sarah K. Croucher and Lindsay Weiss, 1–37. New York, NY: Springer.

Dalley, Cameo, and Paul Memmott. 2010. Domains and the Intercultural: Understanding Aboriginal and Missionary Engagement at the Mornington Island Mission, Gulf of Carpentaria, Australia from 1914 to 1942. *International Journal of Historical Archaeology* 14: 112–135.

Dixon, R.M.W., Bruce Moore, W.S. Ramson and Mandy Thomas. 2006. *Australian Aboriginal Words in English: Their Origin and Meaning*. Oxford, UK: Oxford University Press.

Egloff, Brian, and Wreck Bay Community. 1990. *Wreck Bay: An Aboriginal Fishing Community*. Canberra, ACT: Aboriginal Studies Press.

Fletcher, Daina, and Stephen Gapps. 2012. Nawi: Exploring Australia's Indigenous Watercraft. *Signals* 100: 4–11.

Fowler, Madeline. 2013. Aboriginal Missions and Post-Contact Maritime Archaeology: A South Australian Synthesis. *Journal of the Anthropological Society of South Australia* 37: 73–89.

Fowler, Madeline, Amy Roberts, Jennifer McKinnon, Clem O'Loughlin, and Fred Graham. 2014. 'They Camped Here Always': Archaeologies of Attachment to Seascapes via a Case Study at Wardang Island (Waraldi/Wara-dharldhi), South Australia. *Australasian Historical Archaeology* 32: 14–22.

Fowler, Madeline, Amy Roberts, Fred Graham, Lindsay Sansbury, and Carlo Sansbury. 2015. Seeing Narungga (Aboriginal) Land from the Sea: A Case Study from Point Pearce/Burgiyana, South Australia. *Bulletin of the Australasian Institute for Maritime Archaeology* 39: 60–70.

Fowler, Madeline, Amy Roberts, and Lester-Irabinna Rigney. 2016. The 'Very Stillness of Things': Object Biographies of Sailcloth and Fishing Net from the Point Pearce Aboriginal Mission (Burgiyana) Colonial Archive, South Australia. *World Archaeology* DOI 10.1080/00438243.2016.1195770.

Gara, Tom. 2013. Indigenous Bark Canoes in South Australia. Paper presented at the Flinders University Department of Archaeology Public Seminar, Adelaide, SA.

Garnett, Francis. 1920. Report of the Chief Protector of Aboriginals for the Year Ended June 30, 1920. R.E.E. Rogers, Adelaide, SA.

Garnett, Francis. 1921. Report of the Protector of Aboriginals for the Year Ended June 30, 1921. R.E.E. Rogers, Adelaide, SA.

Gibbs, R.M. 1969. *A History of South Australia: From Colonial Days to the Present*. Blackwood, SA: Southern Heritage.

Gibbs, Martin. 2003. Nebinyan's Songs: An Aboriginal Whaler of South-West Western Australia. *Aboriginal History* 27: 1–15.

Gill, Nicholas, Alistair Paterson and M.J. Kennedy. 2005. 'Do You Want to Delete This?' Hidden Histories and Hidden Landscapes in the Murchison and Davenport Ranges, Northern Territory, Australia. In *The Power of Knowledge, the Resonance of Tradition. Electronic publication of papers from the AIATSIS Indigenous Studies conference, September 2001*, ed. Graeme K. Ward and Adrian Muckle, 125–137. Canberra, ACT: Research Program, Australian Institute of Aboriginal and Torres Strait Islander Studies.

Graham, Doris May, and Cecil Wallace Graham. 1987. *As We've Known It: 1911 to the Present*. Underdale, SA: Aboriginal Studies and Teacher Education Centre.

Griffin, Darren. 2010. Identifying Domination and Resistance through the Spatial Organization of Poonindie Mission, South Australia. *International Journal of Historical Archaeology* 14: 156–169.

Griffiths, Tom (ed.). 1988. *The Life and Adventures of Edward Snell*. North Ryde, NSW: Angus & Robertson.

Heinrich, Rhoda. 1976. *Wide Sails and Wheat Stacks*. Port Victoria, SA: Port Victoria Centenary Committee.

Hickson, Judith. 2012. Bart's Boat. *Goree: Aboriginal & Torres Strait Islander News from the National Museum of Australia* 9(1): 20.

James, Keryn. 2002. Wife or Slave? The Kidnapped Aboriginal Women Workers and Australian Sealing Slavery on Kangaroo Island and Bass Strait Islands. Honour's thesis, Department of Archaeology, Flinders University, Adelaide, SA.

Jeffery, Bill. 2001. Cultural Contact along the Coorong in South Australia. *Bulletin of the Australasian Institute for Maritime Archaeology* 25: 29–38.

Jones, Susanne Montana. 2009. The Anatomy of a Relationship: Doing Archaeology with an Indigenous Community on a Former Mission—A Case Study at Point Pearce, South Australia. Honour's thesis, Department of Archaeology, Flinders University, Adelaide, SA.

Kartinyeri, Doreen. 2002. *Narungga Nation*. Adelaide, SA: Doreen Kartinyeri.

Keating, Claire. 2012. "We Want Men Whose Hearts are … Full of Zeal": An Investigation of Cross-Cultural Engagement within the Weipa Mission Station (1898–1932). Master's thesis, Department of Archaeology, Flinders University, Adelaide, SA.

Krichauff, Skye. 2008. The Narungga and Europeans: Cross-Cultural Relations on Yorke Peninsula in the Nineteenth Century. Master's thesis, School of History and Politics, University of Adelaide, Adelaide, SA.

Krichauff, Skye. 2013. Narungga, the Townspeople and Julius Kuhn: The Establishment and Origins of the Point Pearce Mission, South Australia. *Journal of the Anthropological Society of South Australia* 37: 57–72.

MacKnight, C.C. 1986. Macassans and the Aboriginal Past. *Archaeology in Oceania* 21: 69–75.

Mattingley, Christobel, and Ken Hampton (eds.). 1992. *Survival in Our Own Land: 'Aboriginal' Experiences in 'South Australia' since 1936: Told by Nungas and Others*. Rydalmere, NSW: Hodder & Stoughton.

May, Sally K., Jennifer F. McKinnon, and Jason T. Raupp. 2009. Boats on Bark: An Analysis of Groote Eylandt Aboriginal Bark-Paintings Featuring Macassan *Praus* from the 1948 Arnhem Land Expedition, Northern Territory, Australia. *The International Journal of Nautical Archaeology* 38(2): 369–385.

McCarthy, Michael. 2008. The Australian Contact Shipwrecks Program. In *Strangers on the Shore: Early Coastal Contacts in Australia*, ed. Peter Veth, Peter Sutton, and Margo Neale, 227–236. Canberra, ACT: National Museum of Australia Press.

McGhee, Fred L. 1998. Towards a Postcolonial Nautical Archaeology. http://www.assemblage.group.shef.ac.uk/3/3mcghee.htm. Accessed November 11, 2014.

McLean, M.T. 1933. Report of the Chief Protector of Aboriginals for the Year Ended June 30, 1933. Harrison Weir, Adelaide, SA.

McNiven, Ian J., and Lynette Russell. 2005. *Appropriated Pasts: Indigenous Peoples and the Colonial Culture of Archaeology*. Lanham, MD: AltaMira Press.

McPhee, Ewen. 2001. A Preliminary Examination of the History and Archaeology of the Pearl Shelling Industry in Torres Strait. *Bulletin of the Australasian Institute for Maritime Archaeology* 25: 1–4.

Meide, Chuck. 2013. Economic Relations on a Contested Maritime Landscape: Theoretical Framework and Historical Context. Dissertation position paper no. 1. College of William & Mary, Williamsburg, VA.

Merry, Kay. 2010. Shipwrecks, Castaways and the Coorong Aborigines. In *Something Rich and Strange: Sea Changes, Beaches and the Littoral in the Antipodes*, ed. Susan Hosking, Rick Hosking, Rebecca Pannell, and Nena Bierbaum, 179–194. Kent Town, SA: Wakefield Press.

Middleton, Angela. 2010. Missionization in New Zealand and Australia: A comparison. *International Journal of Historical Archaeology* 14: 170–187.

Mitchell, Scott. 1996. Dugongs and Dugouts, Sharptacks and Shellbacks: Macassan Contact and Aboriginal Marine Hunting on the Cobourg Peninsula, North Western Arnhem Land. *Indo-Pacific Prehistory Association Bulletin* 15: 181–191.

Moody, Stuart M. 2012. *Port Victoria's Ships and Shipwrecks*. Maitland, SA: Stuart Moody.

Morse, Kate. 1988. An Archaeological Survey of Midden Sites near the *Zuytdorp* Wreck, Western Australia. *Bulletin of the Australian Institute for Maritime Archaeology* 12(1): 37–40.

Mullins, Steve. 2012. Company Boats, Sailing Dinghies and Passenger Fish: Fathoming Torres Strait Islander Participation in the Maritime Economy. *Labour History* 103: 39–58.

Mulvaney, John, and Johan Kamminga. 1999. *Prehistory of Australia*. St Leonards, NSW: Allen & Unwin.

Narungga Aboriginal Progress Association. 2006. *Nharangga Warra: Narungga Dictionary*. Maitland, SA: Wakefield Press.

Nash, Michael. 2006. The *Sydney Cove* Shipwreck Survivors Camp. Maritime Archaeology Monograph Series 2, Department of Archaeology, Flinders University, Adelaide, SA.

Newchurch, Ronald, Jr. 2013. Interview by Madeline Fowler. Digital recording. 29 November. Port Victoria, SA.

O'Connor, Sue, and Steve Arrow. 2008. Boat Images in the Rock Art of Northern Australia with Particular Reference to the Kimberley, Western Australia. In *Islands of Inquiry: Colonisation, Seafaring and the Archaeology of Maritime Landscapes*, ed. Geoffrey Clark, Foss Leach, and Sue O'Connor, 397–409. Canberra, ACT: ANU E Press.

Paterson, Alistair G. 2011. Considering Colonialism and Capitalism in Australian Historical Archaeology: Two Case Studies of Culture Contact from the Pastoral Domain. In *The Archaeology of Capitalism in Colonial Contexts: Postcolonial Historical Archaeologies*, ed. Sarah K. Croucher, and Lindsay Weiss, 243–267. New York, NY: Springer.

Reynolds, Henry. 1984. The Breaking of the Great Australian Silence: Aborigines in Australian Historiography 1955–1983. Paper presented at the Trevor Reese Memorial Lecture, University of London, London, UK.

Reynolds, Henry. 2000. *Black Pioneers: How Aboriginal and Island People Helped Build Australia*. Ringwood, VIC: Penguin Books.

Rigney, Lester-Irabinna. 2002. Foreword. In *Alas for the Pelicans! Flinders, Baudin and Beyond*, ed. Anne Chittleborough, Gillian Dooley, Brenda Glover and Rick Hosking, ix–xiv. Kent Town, SA: Wakefield Press.

Rigney, Lester-Irabinna. 2013. Interview by Madeline Fowler. Digital recording. 18 July. Adelaide, SA.

Roberts, David Andrew. 2004. Nautical Themes in the Aboriginal Rock Paintings of Mount Borradaile, Western Arnhem Land. *The Great Circle* 26(1): 19–50.

Roberts, Amy, Jennifer McKinnon, Clem O'Loughlin, Klynton Wanganeen, Lester-Irabinna Rigney, and Madeline Fowler. 2013. Combining Indigenous and Maritime Archaeological Approaches: Experiences and Insights from the '(Re)locating *Narrunga* Project', Yorke Peninsula, South Australia. *Journal of Maritime Archaeology* 8(1): 77–99.

Roberts, Amy, Madeline Fowler, and Tauto Sansbury. 2014. A Report on the Exhibition Entitled 'Children, Boats and 'Hidden Histories': Crayon Drawings by Aboriginal Children at Point Pearce Mission (SA), 1939'. *Bulletin of the Australasian Institute for Maritime Archaeology* 38: 24–30.

Roberts, Amy, Lester-Irabinna Rigney, and Klynton Wanganeen. in prep. *The Butterfish Mob: Narungga Cultural Fishing*. Adelaide, SA: Wakefield Press.

Russell, Lynette. 2005. Kangaroo Island Sealers and Their Descendants: Ethnic and Gender Ambiguities in the Archaeology of a Creolised Community. *Australian Archaeology* 60: 1–5.

Smith, Anita, and Wendy Beck. 2003. The Archaeology of No Man's Land: Indigenous Camps at Corindi Beach, Mid-North Coast New South Wales. *Archaeology of Oceania* 38(1): 66–77.

South, William Garnet. 1915. Report of the Protector of Aborigines for the Year Ended June 30, 1915. R.E.E. Rogers, Adelaide, SA.

South, William Garnet. 1916. Report of the Protector of Aborigines for the Year Ended June 30, 1916. R.E.E. Rogers, Adelaide, SA.

South, William Garnet. 1919. Report of the Chief Protector of Aboriginals for the Year Ended June 30, 1919. R.E.E. Rogers, Adelaide, SA.

Staniforth, Mark, Susan Briggs, and Chris Lewczak. 2001. Archaeology Unearthing the Invisible People: European Women and Children and Aboriginal People at South Australian Shore-Based Whaling Stations. *Mains'l Haul: A Journal of Pacific Maritime History* 36(3): 12–19.

Stanner, William Edward Hanley. 1969. After the Dreaming. Paper presented at the Boyer Lectures, Sydney, NSW.

Sunday Mail. 1974. No title. *Sunday Mail* 27 January. Adelaide, SA.

Taçon, Paul S.C. and Sally K. May. 2013. Special Issue: Maritime Rock Art. *The Great Circle* 35(2).

Taylor, Rebe. 2008. *Unearthed: The Aboriginal Tasmanians of Kangaroo Island*, 2nd ed. Kent Town, SA: Wakefield Press.

Wanganeen, Eileen (ed.). 1987. *Point Pearce: Past and Present*. Underdale, SA: Aboriginal Studies and Teacher Education Centre.

Warrior, Fred, Fran Knight, Sue Anderson, and Adele Pring. 2005. *Ngadjuri: Aboriginal People of the Mid North Region of South Australia*. Prospect Hill, SA: SASOSE Council.

Wesley, Daryl, Jennifer McKinnon, and Jason T. Raupp. 2012. Sails Set in Stone: A Technological Analysis of Non-Indigenous Watercraft Rock Art Paintings in North Western Arnhem Land. *Journal of Maritime Archaeology* 7: 245–269.

Westerdahl, Christer. 1992. The Maritime Cultural Landscape. *The International Journal of Nautical Archaeology* 21(1): 5–14.

Westerdahl, Christer. 2003. Maritime Culture in an Inland Lake? In *Proceedings of the 1st International Conference on Maritime Heritage*, ed. Carlos Alberto Brebbia and Timmy Gambin, 17–26. Malta.

Westerdahl, Christer. 2006. The Relationship between Land Roads and Sea Routes in the Past—Some Reflections. *Deutsches Schiffahrtsarchiv* 29: 59–114.

Westerdahl, Christer. 2008. Boats Apart. Building and Equipping an Iron-Age and Early-Medieval Ship in Northern Europe. *The International Journal of Nautical Archaeology* 37(1): 17–31.

Westerdahl, Christer. 2011. Conclusion: The Maritime Cultural Landscape Revisited. In *The Archaeology of Maritime Landscapes*, ed. Ben Ford, 331–344. New York, NY: Springer.

Westerdahl, Christer. 2014. The Maritime Middle Ages—Past, Present and Future: Some Ideas from a Scandinavian Horizon. *European Journal of Archaeology* 17(1): 120–138.

Wood, Vivienne, and Craig Westell. 1998. *Point Pearce Social History Project, Yorke Peninsula, South Australia*. Parkside, SA: Australian Heritage Commission.

Chapter 5
The Formation of a West African Maritime Seascape: Atlantic Trade, Shipwrecks, and Formation Processes on the Coast of Ghana

Rachel Horlings and Gregory Cook

Introduction

> The study of human cultures cannot be divorced from a study of their environment and the mutual interaction between human activities and environmental processes (Rapp 2000:243).

The entire history of maritime exploration and the resulting international encounters is one of constant movement and change through a range of physical and cultural environments, with the sea as a key player. Interpreting this changing history on the sea and the regions that border it, however, is exceptionally challenging because, as Parker (2001:22) writes, "[the] very basis of sea travel, the surface of the ocean, is changeable and mobile" and therefore cannot preserve *on its surface* any record of the past. Evidence of activities on the sea does, however, remain on the seafloor and is found in literally every area of the world, including in sub-Saharan Africa. As recently as 1998, one researcher predicted that because the environmental conditions along the African coast do not appear to be conducive to the preservation of submerged remains, it "would be worth looking for wrecks underwater [in African waters], but the chances of such a search producing a major source of new information is small" (Unger 1998:224). Recent work suggests that Unger's assumption is overly pessimistic. Investigations of a number of submerged sites in coastal Elmina have provided immense amounts of data; work in the region indicates that there is without a doubt additional cultural material surviving that remains to be investigated (Cook and Spiers 2004; Cook 2012; Horlings 2011, 2012; Pietruszka 2011).

R. Horlings
Syracuse University, Syracuse, NY 13244, USA

G. Cook (✉)
Department of Anthropology, University of West Florida,
11000 University Parkway, Pensacola, FL 32514, USA
e-mail: gcook1@uwf.edu

© Springer International Publishing AG 2017
A. Caporaso (ed.), *Formation Processes of Maritime Archaeological Landscapes*,
When the Land Meets the Sea, DOI 10.1007/978-3-319-48787-8_5

Following the approach taken by many authors in this volume, we feel that the investigation of this submerged heritage is necessarily a holistic endeavor, encompassing historical, cultural, physical, and environmental contexts that combine to create a seascape. The differential preservation and distortion of archaeological sites today is a direct result of both cultural and natural formation processes, and therefore any archaeological investigation must be deciphered and interpreted through the lens of formation processes (Stein and Farrand 1985; Schiffer 1987, 1996; Stein 1987; Murphy 1990; Quine 1995; Reitz et al. 1996; Staski 2000; Delgado and Staniforth 2002). Formation processes in this discussion are understood as the larger, historical circumstances of coastal maritime trade in the Atlantic era in Ghana, as well as those physical processes which served to render heritage submerged and have subsequently played roles in its destruction, preservation, and stabilization. Historical formation processes frame the interpretation of heritage, including a mid-seventeenth century shipwreck known as the Elmina Wreck and other submerged artifacts, in terms of their relationships to the historical trade and its seascape, and physical formation processes dictate the investigation and interpretation of physical remains. Discussions here are approached within the historical and physical seascape context from the perspective of the sea looking onto land. This focus is necessitated by the nature of the data, which at present primarily consist of evidence of European trade and navigation practices located offshore. The seascape itself, however, encompasses both the sea and the adjacent shoreline, incorporating both African and European engagements with each other and with the sea.

The purpose of formation studies is to provide a foundation for the understanding and interpretation of culture and history from the contextualized remains of submerged archaeological sites, emphasizing the "dynamic interactions between humans, the natural environment, and their depositional records" (Goldberg et al. 1993:vii). Formation process research is most helpful when understood as an interdisciplinary endeavor and applied as both a theoretical construct and methodological approach. In doing so, it is possible to present a more complete picture of the processes that created, affected, and continue to impact historical submerged cultural resources, and of their interpretations concerning international maritime interactions of the past Atlantic world. Illustrative of these processes and their archaeological interpretation is maritime research at Elmina, located in coastal Ghana. Elmina's place is at the boundary of land and sea, historically a key juncture in the meeting and straddling of worlds, and a place now providing insights into the intricacies of maritime relationships, navigation within the complex seascape, and varying scales of history and archaeology. While all of the ideas presented here are treated in far greater detail elsewhere (Horlings 2011), this chapter is intended as an introductory discussion to the complex formation processes of the maritime Atlantic trade in West Africa. Historical archaeology has been conducted in Ghana and across West Africa for some time. This specific research was part of the Central Region Project, which combined both terrestrial and maritime archaeological research in studying social, economic, and political transformations during the period of European contact (DeCorse et al. 2000, 2009). The primary focus of this chapter is the maritime component in these transformations.

A Brief History of Trade, Logistics, and the Environment at Elmina

The complex and intersecting historical processes that shaped the Elmina seascape included such factors as the Atlantic trade; international relations between Africans and Europeans or the foreigners trading at the coast; permanent presence of Europeans at Elmina Castle; resources of the Benya Lagoon; common sailing, navigational, and anchoring practices for international vessels; the central location of Elmina both in terms of the Ghanaian cost and of the trade along the West African coast as a whole; and the weather, and oceanographic patterns that dictated seasons of trade. The formation processes that determined the creation, destruction, and preservation of submerged cultural resources included all these factors; the disparate physical forces that affected historical ships and sailors now affect the tangible remains of their presence across the submerged landscape at the boundaries of land and sea, representing the intersection of all these processes. A brief look at the different historical and physical formation processes of the region provides the foundation for interpretation of cultural heritage within the historical maritime seascape.

Historical Formation Processes

Originally called *São Jorge da Mina*, Elmina Castle was established by the Portuguese in 1482 and was the first European trading post in sub-Saharan West Africa, built in an effort to protect the Portuguese gold trading interests (Fig. 5.1). In 1637, Elmina Castle was captured by the Dutch and remained in Dutch hands as a trading establishment until it was ceded to the British in 1872 (Feinberg 1989; Ballong-Wen-Mewuda 1993; DeCorse 2001; Yarak 2003). Not only was it the first permanent European establishment in sub-Saharan West Africa, but the town and the trading fort of Elmina was considered by many to be the most important trading center on the Ghanaian coast for at least two centuries (Baesjou 1988:49–50; de Marees 1987:218–222; Feinberg 1989:v, 2; den Heijer 2003:149; Van Den Boogaart 1992:373; DeCourse 2010). The early establishment and continuous use of the permanent trading location on the coast are particularly important in terms of understanding the potential for shipwrecks and other submerged cultural resources in the area, as the more vessels that visited, and the long duration of maritime trade

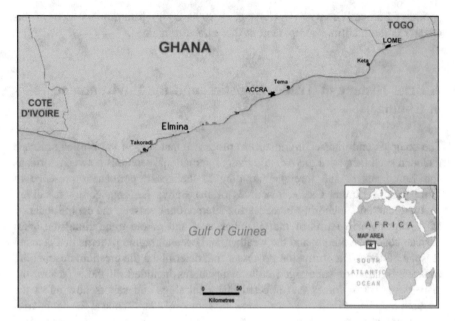

Fig. 5.1 The town of Elmina is located in central coastal Ghana (Horlings 2011:8)

at this location, increased the likelihood of wrecking, misadventures, and the creation/destruction of other submerged cultural remains.

The principle advantages in the location of the castle were that it could be easily defended, had access to fresh water, and already-established local networks of trade and resources (Feinberg 1989:41–42), and had a bay and beach area in which to careen, conduct trade, and protect small vessels and goods against storms (Lawrence 1963:103; da Mota and Hair 1988:61; Hair 1994:16–17) (Fig. 5.2). In addition, the Elmina promontory is made up of sandstone and other rock that provided adequate building materials for the construction of the fort, and the heavy seas and large swells/waves breaking over the rocks (in all seasons) south of the castle provided a defense against potential enemies (US Navy Hydrographic Office 1951:52; Lawrence 1963:103; Ballong-Wen-Mewuda 1993:73).

Elmina's location in the pseudo-harbor area of the Benya River and shallow Elmina Bay was all the more important because, with few exceptions, access to the coast of much of West Africa is difficult due to the lack of large river mouths and natural harbors, and the abundance of high waves (Zook 1919:183; Bourret 1949:8; Dickson 1965; Allersma and Tilmans 1993:235; Rawley and Behrendt 2005:10). This, in addition to the inherent dangers of sailing and navigation (Martin 1837:228; Adams 1966[1823]:239, 243; Thomas 1969[1860]:193; Brooks 1970:235; Smith 1970:516–517; Feinberg 1989:67–68; Mitchell 2005:178–180; Barbot's comments in Hair et al. 1992:382; Mitchell 2005:178–180), dictated that merchant vessels would sail and anchor off the mouths of rivers or as far offshore as necessary to avoid dangers such as reefs or shoals. They generally anchored in

Fig. 5.2 Overview of the Elmina peninsula and Elmina Bay as they exist today. The approximated nineteenth century shoreline is outlined in *red*. *Note* the modern harbor entrance to the east of the Castle and the Benya Lagoon running roughly parallel to the coast (Horlings 2011:82)

seven or eight fathoms of water if possible, depending on the vessel's draught, but even up to 24 fathoms depth, equating to three or four miles offshore (Bold 1823:56). Vessels anchoring in these roadsteads used ships' boats or local canoes to transport people and goods from ship to shore (Gutkind 1989; Thomas 1969 [1860]:193; Brooks 1970:235; Smith 1970:516–517; Feinberg 1989:68; Hair et al. 1992:382). The unpredictable seafloor topography and currents required constant vigilance for vessels anchored offshore (US Navy Hydrographic Office 1951:10).

Where inconsistencies in historical accounts of use and anchoring in and near the Benya River and Elmina Harbor area are problematic in terms of reconstructing historical trading and anchoring practices, they are useful in highlighting some of the historical uses of the area and changes made over time to the maritime infrastructure, for instance, modifications to the peninsula itself (DeCorse 2001). Despite the fact that most of the accounts we have were not written by sailors, they can nevertheless provide glimpses into how the maritime world was accessed, understood, and manipulated over time (see Horlings 2011:80–91 for a more complete discussion). The changing physical environment of the Castle and Lagoon area, including intentional modifications of the lagoon and the peninsula as well as natural changes such as the dramatic sedimentation in the area, in concert with changes in ship technology and sizes, likely dictated what activities were plausible at different times. Archaeological investigations are starting to shed light on these aspects of historical maritime trade in West Africa.

Once trade was regularly established between Africans and Europeans on the Ghanaian coast, it accelerated rapidly. Between the end of the fifteenth century and the third quarter of the eighteenth, more than 50 European posts were built in the area now called Ghana (van Dantzig 1980; Posnansky and DeCorse 1986; DeCorse 2001, 2010), with *São Jorge da Mina*, or Elmina being the first. As the Atlantic trade burgeoned into a truly global affair, various European nations rose to economic and trading power, competition and manipulation grew between nations and between private traders, smugglers and pirates challenged state enterprises, and fierce confrontations for control of trade monopolies were fueled (Bourret 1949:16; Nørregård 1966; Bandinel 1968 [1842]:49; Calvocoressi 1968; Ellis 1969 [1893]:29–38; Brooks 1970:4; Crooks 1973 [1923]; Bean 1974; Blake 1977; de Marees 1987:188; da Mota and Hair da 1988; Ballong-Wen-Mewdua 1993:457–459; Rømer 2000; Ward 2003; Mancke 2004:161). Also during this time, trading posts and depots were established, captured, and recaptured; vessels were captured or destroyed; and difficulties in communication and conflicts of interest created endless confusions between and within different nations involved in this trade (Nathan 1904; Hopkins 1973:92; Thornton 1992:108; Eltis et al. 1999; Schwartz and Postma 2003:176). It was within these historical formation processes that maritime trade occurred at Elmina and within these processes that maritime cultural heritage in now interpreted and understood.

Physical Formation Processes

The physical environment played a significant role in the shaping and expansion of maritime trade. It affected trade patterns, provided the arena within which the Atlantic trade in West Africa was enacted, and continues to play a dramatic role in the destruction and preservation of remaining traces of historical trade. This environment, including the sea and its attendant shore region, both past and present, is an intrinsic part of the formation processes that created and affected submerged sites. As a result, a basic familiarity with its various features is essential for understanding sites found within this context. In particular, the sea constituted a real space that was recognized as an entity in and of itself (Steinberg 2001:109), in many way dictating the lives of those who depended on it. The sea and its shore have played two distinct roles in the formation of archaeological sites; these include dictating use and navigation of the region, as well as transformations of submerged cultural heritage. A brief summary of weather, sea conditions, and physical geography—all factors affecting sailors in the past—serves to illustrate the first, and many of those factors still active on submerged sites today characterize the second. A discussion of these complex factors is useful in illustrating their formation process roles.

For purposes here, site formation processes are defined as the historical, physical, chemical, and biological processes, which create and transform archaeological sites through time. These include currents, sedimentation, storms, benthic organisms, the events that occurred to cause items to be submerged, such as a shipwrecking event, and any subsequent interactions with submerged sites. Formation process studies examine the relationship between long-term archaeological and environmental trends, both at the macro- and micro-level, as a means of identifying and accounting for the biases inherent in the creation and interpretation of archaeological sites (Gladfelter 1977; O'Shea 2002:212; Ward and Larcombe 2003:1223). Additionally, because they are active on the levels of artifact, site, and region (Schiffer 1987; Stewart 1999:578; Staski 2000:43), they provide a means of looking at data across a range of scales and set the stage for the interpretation of submerged cultural heritage relating to individual events, such as a single shipwreck, or to processes, such as the development of trade logistics, relating to the historic Atlantic trade (Horlings 2011).

Weather dictated both when the Ghanaian coast could be navigated as well as how it was navigated, particularly in the near-shore and roadstead zones. Physical challenges to sailors in this region included restricted visibility from heat refraction and thick, dry season dust that often rendered visibility onshore and offshore essentially zero, resulting in problems taking sights and making approach to land particularly dangerous (i.e., Bold 1823:39–42; US Navy Hydrographic Office 1951:16). Extreme temperature and humidity changes posed challenges for sailors having to adapt to new conditions, and for the integrity of wooden vessels in terms of drying and wetting of the wood (Bold 1823:42). In addition, because of the often violent weather in the rainy season, optimal sailing to, from, and along the coast of West Africa was from September through April during the (mostly) dry season (Adams 1966 [1823]:164; Hopkins 1973:107; Awosika et al. 1993:31; Opoku-Ankomah and Cordery 1994:552; Gu and Adler 2004:3366). Although weather was/is never completely predictable, winds were fairly consistent during this time, which favored sailing (US Navy Hydrographic Office 1951:17) and also facilitated shipboard trading due to calmer seas and near-shore conditions.

While trading logistics were a challenge in the dry season, they were far more so in the rainy season. In fact, when possible, vessels generally avoided the west coast of Africa between May and September, but this was not always possible as, for instance, they could be caught in storms, in the doldrums, meet contrary winds, or be held up in trade, delaying schedules and likely resulting in a rainy season stay. In addition to other problems, these delays had potentially serious deleterious effects on ships, as the warm waters of the coast host numerous wood-boring mollusks and other organisms that made the wood of the vessel extremely susceptible to disintegration and other structural concerns. Toward the end of the dry season, storms, and squalls began to hammer the coast regularly and severely, posing dangers both at sea and in attempting to transition between ship and shore and putting restrictions on trading during that period (Bold 1823:40–55). Seasonal rainstorms also posed serious issues to those not at anchor, as sailing in the constantly heavy and difficult seas increased the already high chances of encountering trouble, and possibly wrecking. In addition, sitting for long periods of time also meant greater danger of being surprised and attacked by competitors (Vogt 1979:15–16).

Travel in any season was often difficult and erratic on the coast, owing to hidden rocks and changing sedimentation rates, as well as weather and strong currents. For instance, de Marees (1987:86) comments that it would take only 24 h for the Dutch to sail between Cape Coast (Cabo Corso) and Mori (Mourre) (a distance of less than 10 km), but it could take three or four weeks to make the return journey because of the powerful east current (van Dantzig 1980:17–18; Horlings 2011:77). Like ancient mariners, sailing along the east coast for Europeans (and presumably the Africans as well) depended heavily on the recognition of coastal marks until well into the nineteenth century and later (Bold 1823). The lack of charts and other navigational that aids along the West African coast was the result of most areas not having been surveyed or being poorly recorded. There was also a tendency for different nations and individual traders to keep information to themselves as a means of preventing others from successfully trading in the region (Bold 1823:1–1v). The combination of weather, unfamiliar territory, and the reliance on visual landmarks meant that navigators could be easily misled and their ships endangered along the dangerous coastline (Bold 1823:ii; Cook and Spiers 2004:18). While the majority of vessels sailing and trading along the West African coast survived their expeditions in this often harsh environment, frequently returning to trade again, a significant number succumbed to the natural and manmade dangers prevalent within the Atlantic trade seascape. Although Ghana's coast may have been suitable for building forts and trade networks (van Dantzig 1980:ix–x), it was not a kind place for ships. These forces, combined with the political and economic struggles constantly at play, made the work of the sailor/merchant challenging indeed.

Shipwrecking, capsizing of ships' boats and canoes, anchor or equipment loss, and a host of other events were responsible for the deposition of cultural material on the seafloor, initiating the physical formation processes of cultural heritage. Many of the same factors affecting sailing and trade—currents, storms, sedimentation, sea life, use of the sea—continue to affect submerged heritage today. Cultural heritage from this era already discovered provides clues both to the history of the seascape, and to the subsequent site formation processes at the level of the seafloor. The discovery of additional heritage will only serve to provide a fuller and more vivid perspective on both the historical and formation processes of this region and pivotal period in history.

The Seascape: Sea–Land Interface

Human lives are lived on surfaces, but archaeological surfaces are not necessarily those on which the archaeologist walks (Dincauze 2000:195).

The sea was the connection between the disparate places of the Atlantic world, and the seascape was the theater in which historical maritime interactions took place, and which incorporated the physical, social, economic, and cultural inter-actions and embodiments of trade. It was primarily from the sea that European traders viewed the international trade and the people with whom they interacted (Meredith 1967 [1812]:17, 35; Curtin 1984:57; Baesjou 1988:5, 32, 45; Rømer 2000:16, 27), and it was largely toward the sea that Africans on the coast viewed the same. These places of cultural mingling and interface, with their attendant interactions on the water, whether in harbor or offshore, and onshore, formed vital parts of the multifarious historical seascape and are integral to the study of historical maritime trade today (Westerdahl 1992; Anuskiewicz 1998:224; Lenihan and Murphy 1998:237; Dellino and Endere 2001:219; Parker 2001; Breen and Lane 2003; Cooney 2003; Ash 2007; Chapman and Chapman 2005; Flatman et al. 2005; Dellino-Musgrave 2006:27).

By harnessing the sea, the sailing vessel revolutionized trade and international relations, and it was in great part because of this instrument that continents, including most of Africa and Europe, could meet. In essence, the ship, in its contextual setting of the seascape, (which here includes everything from the Atlantic Ocean to the mouth of the Benya River in Elmina, to the trade establishments onshore), was the instrument that blurred continental boundaries and restructured "our mental geographies of space, place, time, culture, and history" (Ogundiran and Falola 2007:35). But ships and sea could not sustain trade alone—it was the connection to the shore, with its attendant peoples, trade facilities and institutions, and resources that made it so important for the intersection of vastly different worlds. Trade establishments such as Elmina Castle, along with their roadsteads, careening areas, harbor locations, and essential resources such as water and food, provided transition points between the sea and land for both people and goods. Without the established trade networks and infrastructure afforded by these things, maritime trade of any description could not have developed into the global phenomenon that it has.

For all of those involved in historical maritime trade, whether viewing the sea or the land, a working knowledge of all of the components of the seascape was vital to engaging with and assessing international relations, trade opportunities, and dangers posed by humans and nature. Position within the historical seascape often dictated the boundaries between people, but it also served to facilitate the crossing of boundaries between sea and shore, merchant and trader, sailor and land dweller. To the maritime archaeologist, every aspect of this seascape is a vital component in understanding peoples of the past (Cunliffe 2001).

Mutual Formations of Seascape and Trade

Even working with only bits and pieces of the often-obscure material record of the past, we need to have some way of trying to understand how people "engaged with one another across space, how they chose to manipulate their surroundings" (David and Thomas 2008:38), and how their actions were ordered by their surroundings and environment. Aspects of landscape archaeology provide a flexible, or "usefully ambiguous" (Gosden and Head 1994:113; Branton 2009:51), means of incorporating both the physical and social worlds of the past that can be used for the interpretation of these bits and pieces and their contexts (Crumley and Marquardt 1990; Firth 1995; Hood 1996:121; Knapp and Ashmore 1999:2). It also provides a means of "establishing a more integrated understanding of past coastal landscapes or seascapes" (Breen and Lane 2003:469) and a meaningful way of integrating environments of historic interactions with events that created submerged archaeological sites and with archaeological materials (Gosden and Head 1994:115). It is impossible to understand international maritime interactions, cultures, and the archaeology of such, apart from the marine environment, seascape, or cultural maritime landscape in which they existed and exist today (Semple 1931:59; Murphy 1983:85; Westerdahl 1992:5; McErlean et al. 2003:2; Breen and Lane 2003:479; Cooney 2003:325; McConkey and McErlean 2007), and this is especially true in the maritime Atlantic trade in Africa.

Kelly and Norman (2007:173) write that landscape archaeology in Atlantic Africa "addresses the social creation of the landscape within the complexities of the localities that were nested within global systems of distribution and change ..." DeCorse (2001), Richard (2006), and Stahl (2001) provide further discussion of the Atlantic Africa diaspora. The concern here is not only in the social *creation* of the landscape of maritime trade, although it certainly is a significant factor, but also in the ways that the landscape *shaped* social interactions and affected all of the actors "nested" in the trade (Branton 2009:62). Humans conceptualize and modify many of these spaces in culturally specific ways in an interactive process in which people both condition and are conditioned by the environment or landscape (Gosden and Head 1994:113; Hood 1996:139; David and Thomas 2008). This mutual shaping of seascape and trade then provides the context for understanding submerged sites (Breen and Forsythe 2001) and views the past from outside the boundary of sea and land, because it encompasses both (Morgan and Green 2009:12–13), including the blurred boundaries between peoples who met and lived and interacted and traded within the seascape setting.

The historical seascape was "actively inhabited space" (Knapp and Ashmore 1999:8) used for navigation, anchorages, sources of fresh water, food supplies, and people with whom to trade, and it dictated methods of trade. This understanding of the seascape does not blur the distinction between the maritime and terrestrial realms (McCarthy 2006:7) in coastal Ghana; rather, it demonstrates their interconnectedness, encompassing also submerged cultural resources that occur at and across the boundaries of the physical and social land/seascape. The boundary between sea and land is not arbitrary (cf. Crumley and Marquardt 1990:74), it is concrete and real, and yet ever-changing and unpredictable. The social consequences of its presence are highly contextual, yet have some of the same effects on every culture and interchange that occurs along the boundary. The boundary of sea and shore is integral to an understanding of the seascape and is an integral part of these "fields of human engagement" (Head 2008:379).

What more appropriate description could there be of the historical seascape in coastal Ghana? The ocean, foreshore, and shore made up the arena upon which contacts, trade, wars, rivalries, and communications were staged, and within which international relations were engaged during the time of the Atlantic trade (Russell-Wood 2009). But there were physical, social, and cultural boundaries to these engagements as well (Bower 1986; Marquardt and Crumley 1987; Preucel and Meskell 2009:220). For Europeans and Africans alike, the maritime trading space became a new frontier (Morgan and Greene 2009:12). While Europeans had been traversing vast areas of the ocean coasts for centuries (Cunliffe 2001:37–39), operating in West Africa was for some time a new frontier. Africans/people of the Ghanaian coast had for a long time been fishing in the ocean and conducting lateral trade along the coast (Smith 1970; Feinberg 1989:5, 19; DeCorse 2001:108; Morgan 2009:223), but venturing out into the ocean, and essentially *across* oceans, was not at all practiced in most of Africa, and indeed was still relatively new for most of the world (Parker 2001:38; Mitchell 2005:178–179; Chaplin 2009).

Within this seascape environment, the Europeans were constrained by the cultural and physical boundaries of the rough coastal waters (DeCorse 2001:17), and were not able to bring their large sailing vessels in direct contact with the African world. In addition, cultural and political restraints put in place by local Africans dictated access to shore and to trade (Meredith 1967 [1812]:103–104; Priestley 1969:8; Hopkins 1973:108–109; van Dantzig 1980; Curtin 1984:15–16; de Marees 1987:81, 207; Feinberg 1989:vi, 2; Thornton 1992:7; Ballong-Wen-Mewuda 1993:444; Mancke 2004:149–150; Mitchell 2005:202; Pearson 2008:16, 22–23; Morgan 2009:225). And yet, with the help of local peoples who allowed and assisted them (using canoes) to come to shore, they traversed physical, cultural, and social boundaries (Pearson 2008:22). Likewise, the Africans who engaged with the Europeans on their vessels were given snapshots across cultural and space boundaries into worlds mostly alien to them (Feinberg 1989:vii). These exchanges, part of what Kelly (2004:225) terms the "mutual discovery of Europeans and Africans," represented the blurring of the social and physical boundaries of two completely different worlds (Marquardt and Crumley 1987:8–9; Ward 2003; Yarak 2003; Johnson 2007:15). And it was within the constraints of the sea, within the context of these boundaries and frontiers, that vessels struggled and sank (Adams 2001:292; McCarthy 2008), leaving a concrete record of both physical and ephemeral events of the past.

The seascape is primarily, therefore, the medium for action—for trade, social interactions, conflicts, and exploration. The focal point on this stage of history is Elmina, straddling the boundary of land and sea. The location of Elmina Castle on shore was a place to which merchants came, but the coastal zone, the ocean, and the surrounding lands, made up the larger landscape and seascape within which historical maritime trade actually took place (cf. Torrence 2002:766; Branton 2009:52; Preucel and Meskell 2009:215). The roadstead passing in front of Elmina was one of the most trafficked zones along the West African coast due to Elmina's trading prominence [until the destruction of the settlement in 1873 (DeCorse 2001:30)] and its central location on the coast. In addition, the constant longshore (Guinea) current running from West to East served as the most efficient conduit for sea travel along the coast, bringing vessels often within sight of the castle. Citing codices in the Lisbon archives, Vogt (1979:166) records that in the early seventeenth century, the

Dutch alone had about 60 vessels a year on the Elmina coast, an estimate that may in fact be quite low (Coombs 1963:2; Dickson 1965:103). The Elmina seascape fostered this booming trade, and in turn played a dramatic role in the history of the Atlantic trade.

In addition to the sea and roadstead, Elmina's strategic position on the Benya Lagoon offered the rare luxury along the West African coast of some protection from the open sea, a shoreline arena for trade, and even a careening point for some vessels (Feinberg 1989:27; Hair 1994:16; DeCorse 2010). The history of the use of the Benya Lagoon and adjoining Elmina Bay areas is complicated and not yet clearly understood, but nevertheless is an important part of the history of the region (Horlings 2011:86–101). Patterns in both the historical and archaeological records concerning the Benya Lagoon and Elmina peninsula (DeCorse 2001:52–53) suggest that there have been dramatic changes in patterns of use of these areas over the centuries (Hair et al. 1992:380), and provide some insights into local peoples' use of them. Discovery of the remains of an early eighteenth century, likely Dutch, vessel in the Benya Lagoon (Pietruszka 2011) attests to the use of this conduit between land and sea, but to date relatively little research has been done pertaining to the European use of this region. The meeting of land and sea in the Elmina seascape, and the formation processes associated with it, offer untold possibilities for archaeological investigation.

The strength of maritime archaeology lies in its ability to traverse geographic and other boundaries (McGhee 2007:393), investigating the local with the global (Ogundiran and Falola 2007:41), and the connections and influences of international relations such as those between Africans and Europeans who met in the contact zones at the edge of the African seascape (DeCorse 2001:10; Mitchell 2005:173; Dellino-Musgrave 2006:28). The potential for understanding many of these connections through the physical remains of those interactions in the seafloor is just beginning to be realized for much of Africa, and for West Africa it is truly in its formative years. Integral to archaeological interpretation, then, is an understanding of scale and spatial patterns and processes with the underwater landscape and its relation to the historical seascape above it (Garrison 1998; Harris 2006:50).

The Underwater Seascape and Environment

The need to know the seafloor as well as the conditions on the surface of the sea is something that sailors have understood since the early days of water travel. Parker (2001:33) writes that since ancient times, sailors have tested the seafloor to sample bottom conditions, in effect creating for themselves a map of the underwater landscape of the seafloor. This knowledge of the underwater landscape provided them with indications of their positions and environment and provided additional means, albeit often indirect, by which they could "articulate their voyage" (Parker 2001:33). While historical sailors on the west coast of Africa were by necessity familiar with sea bottom conditions along the coast, their pictures of the underwater landscape were clearly different than those available through the technological tools and advances available to sailors and researchers today. A basic description of the underwater landscape around coastal Elmina will provide not only a picture of the physical setting of submerged cultural resources, but will also highlight local formation process agents that have served to destroy and preserve the underwater resources (Horlings 2011).

Three basic seafloor characteristics, found singly or in various combinations, may for the most part be seen in the side scan sonar data, and are corroborated by the literature and diver observations (John et al. 1977; Koranteng 2001:2). These include areas consisting of rock reefs and low relief sandstone formations; large expanses of sand, some of it formed into ripple patches of various sizes and shapes; and a few, relatively small and isolated patches of distinct black, sticky mud (i.e., Koranteng 2001:2) that have been observed by divers, but appear relatively featureless and having very low backscatter returns in the side scan sonar signatures. Cultural materials were discovered in rocky reef, sand, and mixed sediment areas. Comparisons between the side scan sonar data collected 6 years apart (2003 and 2009) demonstrate both consistency and some variation in the different geomorphological seafloor features, likely the result of relatively consistent interactions between oceanographic agents and local geomorphological features.

The interrelated systems responsible for the movement, transport, erosion, and deposition of sediments on the continental shelf are extremely complex (Weggel 1972:2). Some of these are highlighted here, as it is important to note the significance of these formation agents in distributing and modifying sediments, and therefore in both the underwater seascape and on submerged archaeological heritage. Three primary agents include currents, upwelling/downwelling, and storm surge. The coast of Ghana is characterized by high (often exceeding 1 m) waves (Amlolo 2006) and powerful currents. The primary current along the coast is the Guinea Current, which travels from West to East, approximately parallel with the coast, and varying in speed from less than 1 kt to more than three (White 1989:160), and served as the ship highway of the Atlantic trade. Driven by offshore winds and the Guinea Current, waves strike the coast at an oblique angle, thus producing the longshore drift; the action of the longshore and littoral drift are responsible for the predominate and most powerful movement of sediments from West to East, parallel with the coast near Elmina. While the movement of sediments has dramatic effect on the shoreline, as well as on economic activities (Horlings 2012), it is also enormously important in terms of submerged heritage, as sediments deposited or removed from archaeological sites in the course of one or several events, or over years are critically important for their destruction and preservation. Changes in environmental agents or processes are recorded in the sedimentary record, indicating both events and processes in time and across space (Stein 1987), and this sedimentary history provides insights into the formation processes affecting submerged heritage.

The event of the shipwreck or loss of other cultural material to the seascape ties together the larger historical setting and context of trade, and provides the tangible remains that may be studies, queried, and investigated (Adams 2001:299; Gould 2001:195; Green 2004). These relationships are crucial, because as evidence is interpreted, the social and historical implications concerning the archaeological heritage and its ties with the larger historical and physical/environmental contexts come full circle, providing insights into historical maritime trade. The nature of archaeological material evidence is that it is inherently generated through a series of timescales; it is built from a single event, such as a shipwreck or loss of an anchor, and from the results of processes that occur and build up over centuries or millennia, such as the development of trade forts and evidence of the navigational practices associated with them, through a complex combination of human and natural forces (Gosden and Kirsanow 2006:30). This material is then part of the "profoundly contextual" (van Dommelen 1999:283) seascape of coastal Elmina, articulating with what Preucel and Meskell (2009:216) call the "politics of location and the social construction of space and place" as it relates to the overarching historical formation processes of the region.

The Archaeological Record Bears Witness

Historical archives are filled with fascinating tales of maritime adventures and woe, and it is a fortunate researcher indeed who can with certainty relate such a tale to archaeological remains discovered in some corner of the sea. Historical accounts give life to these inanimate remains, and provide a means of relating a far-distant past with tangibles of the present; they connect people across the ages by relating human experience. In addition, whether they can be directly associated with archaeological remains or not, accounts that are unquestionably part of the same historical seascape in which archaeological remains are found to highlight the stories of humanity that touch all of the past and present. It is these same accounts, however, that are largely lacking in terms of the maritime history of coastal Ghana, and as such, it is all the more important to find and investigate the archaeological heritage as a means of telling parts of the story of the past of this globally important and yet little understood facet of maritime history.

Shipwrecks and other submerged cultural sites represent more than just culture, history, and bits of broken things (Flatman 2003); they also are the embodiment of different scales—scales of time, culture, and land and seascapes of formation processes, and events. The scale of an event can relate to the entire Atlantic trade (Gilchrist 2005:331–332; Morgan and Greene 2009:7), or to a single wrecking, and formation processes can do the same (Wandsnider 1998:87; Veth 2006:23). It is at the intersection of these varying scales and processes that the value of shipwrecks in understanding past interactions lies; it is there that we see both the individual vessel and its place in the overarching worlds and dramas of which it was a part, culminating in its wrecking (Foxhall 2000; Staniforth 2003). By beginning with the extant physical remains, it is possible to explore and understand the importance of the original vessel at different scales (from single artifacts to entire shipwrecks), investigated together with events, times, processes, and the historical seascapes and landscapes of maritime trade. It is through the investigation of shipwrecks and other submerged sites, their environs, and the processes that affect them that we can begin to glimpse the history from, and assign meaning to, the tangible bits of broken things we find at the bottom of the sea (Firth 1995; Gould 2001:197; Staniforth 2003:29; Flatman 2003).

All of this can be illustrated by a shipwreck site, changes in the land–sea interface, and several other submerged heritage sites in coastal Ghana. Contextualizing these is accomplished by approaching them as part of the larger whole of historical maritime trade in Africa and as integral parts of the past and present physical environment of coastal Elmina. It should be noted that while modern populations are certainly having effects on submerged heritage, the lack of diving technology on the coast has to date preserved sites from intentional disturbance; while it is unknown how long these protections will remain (Horlings 2012), for purposes here, modern influences on the sites are not discussed as major formation process factors.

The Elmina Shipwreck

The Elmina Wreck site is located approximately 2 km southeast of the Elmina peninsula and is in 11–12 m of water in a highly dynamic ocean environment (Fig. 5.3) (Anthony and Blivi 1999:165). Discovery of the Elmina Wreck was

Fig. 5.3 View of the approximate historical roadstead/anchorage off of Elmina (indicated by *red dashed lines*) and the location of the Elmina Wreck (Horlings 2011:84)

through a side scan sonar remote sensing survey conducted in 2003, and three field seasons on the site resulted in a site plan produced almost entirely in zero-visibility waters (2005), limited excavation and extensive sediment core sampling (2007), and additional remote sensing and monitoring the site (2009) (Cook and Spiers 2004; Horlings 2011; Horlings et al. 2011; Pietruszka 2011; Cook 2012). Exposed portions of the site consist primarily of six cannon, an anchor, and trade items consisting of stacks of brass and pewter basins, barrels of brass manilas, lead rolls (which may or may not have been for trade), trade beads and cowry shells scattered throughout the site, indications of cloth cargo, and a number of other objects, including the remains of a birch tree that was likely live cargo on the vessel (Fig. 5.4). The site is underlain by a solid concretion throughout most of it. In addition, the presence of wood from the hull was confirmed through collection of sediment core samples. Dramatic sedimentation on the site, monitored over the period of six years since its discovery, is illustrated by the fact that in 2009, more than two-thirds of the site was completely covered by sediments, dramatically changing its surface signature (Fig. 5.5). It is anticipated that the site is currently completely invisible on the surface due to sediment coverage, a significant factor to consider in the search for other submerged archaeological heritage in the region.

Investigations indicated extensive mixing across the site, and artifactual material within it spans a period of at least 400 years, from the sixteenth century through the twenty-first, making interpretation based on artifacts extremely difficult. Research indicates, however, that it is likely a Dutch vessel of the seventeenth or eighteenth centuries (Pietruszka 2011; Cook 2012), and radiocarbon dates taken from wood collected in sediment cores confirms the mid-seventeenth century date (Horlings 2011:264–265). A possible candidate for the identity of the vessel, a Dutch merchantman named *Groeningen*, which sank near Elmina in 1647, has been proposed based on historical research (Furley Collection Notebook 1646–1647 pages 162–164, transcribed by C.R. DeCorse, August 2010), but this cannot be confirmed until more research is conducted.

Formation process research at the site, based primarily on material recovered in sediment cores (Horlings 2009, 2013), as well as assessment of the macro remains at the site, has provided insights into a range of topics, including causes of sinking and the vessel's subsequent tenure on the seafloor (Horlings 2011:293–294). A brief summary of the results of these investigations reads as follows:

> Sometime in the dry season of the mid-17th century, a European trading vessel caught fire belowdecks. It sank on a relatively even keel, with decks just below the surface of the water, possibly providing an opportunity for limited salvage. Over time the wreck was colonized by various biota, but the combined pressures of an active and dynamic seafloor environment and marine boring organisms caused the vessel to eventually collapse in on itself, preserving original lading even as the hull disintegrated. Cycles of burial and exposure served to trap and mix the original materials of the vessel with those of other centuries that were also in those waters. When discovered in 2003, the vessel once again became part of the known seascape of coastal Ghana, and an active part of maritime history there.

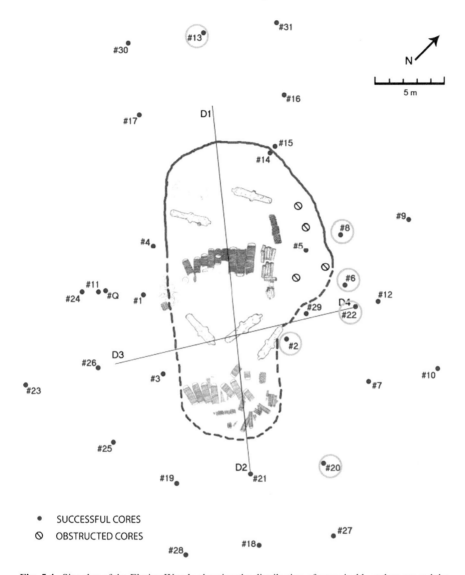

Fig. 5.4 Site plan of the Elmina Wreck, showing the distribution of cores in *blue* taken around the site. Solid *red lines* indicate verified extent of concreted mass under trade goods. *Dashed lines* show estimated extents of concreted material. *Green circles* indicate where wood from the vessel's hull was collected in cores

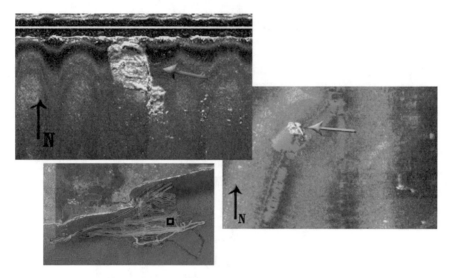

Fig. 5.5 Side-scan sonar mosaic off of Elmina, with details of the wreck site from 2003 (*left*) and 2009 (*right*). Although apparent as a sonar anomaly in both seasons, the wreck is clearly more exposed in the 2003 image, and the 2009 image suggests a substantial portion of the southern extent is covered by soft sediments (Horlings 2011:231)

The Benya Lagoon Vessel

The single archaeological example presently known of European use of the Benya Lagoon for shipping-related activities may be found in the remains of a vessel, likely a hulk, that were discovered by dredgers in 2007 in the western or upper part of the lagoon (Fig. 5.6) (Pietruszka 2011). Found in material dredged from 3 m below current sediment levels, most of the remains were broken and destroyed by the time archaeologists were called to investigate them. A total of 15 various hull elements and three cannon (originally four, but one disappeared, likely taken for scrap metal) comprise the remains and make a thorough assessment difficult. Dendrochronological dating of timbers provides a date of the turn of the eighteenth century and a provenience of northern Europe; these clues, in addition to analysis of the cannon and historical documents, indicate a likely interpretation of the vessel having been Dutch (Pietruszka 2011). While a number of theories have been proposed, it is presently unknown what its purpose was so far up the lagoon (Horlings 2011; Pietruszka 2011), and since the site has now been likely completely obliterated (Horlings 2012), it is unlikely that a great deal more will be known from this vessel. Its presence in this intermediate zone between land and sea, however, provides another piece to the puzzle of understanding the historical maritime past of the coastal Ghanaian seascape.

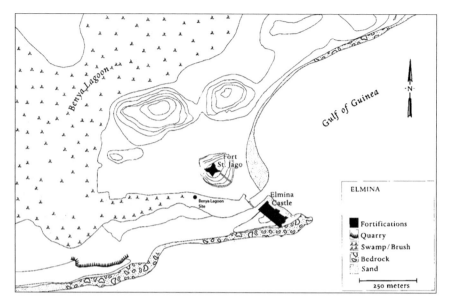

Fig. 5.6 Map showing the location of the Benya Lagoon site (Pietruszka 2011:175, adapted from DeCorse 2001:53, Fig. 2.6)

Other Submerged Heritage

Space does not allow a full treatment of the remaining known cultural heritage in the Elmina seascape, but a brief discussion will highlight both the variety of heritage currently known to be present, and the potential it provides for investing the physical and cultural maritime seascape.

High sedimentation rates in the region are likely responsible for the difficulty researchers had in locating and identifying several targets in the 2003 side scan sonar data, as exemplified by the Single Anchor Site, which was noted as a target in the 2003 side scan data, dived multiple times with negative results in 2007, noted once again in the 2009 side scan sonar data, and discovered in 2009. The site consists of a single anchor that appears to have been set, and is located approximately 400 m ESE from the Elmina Castle, just inshore of a large, natural sandbar at the southern send of the Elmina peninsula and in approximately 5 m water depth. Only a single fluke was visible above the sediments; it measured approximately 50 cm across, suggesting that the total size of the anchor is likely between 3 and 4 m in length and possibly a British Admiralty anchor of the eighteenth or possibly early nineteenth century (Cotsell 1858:14–17; Rubin 1971:237; Curryer 1999:49). Interestingly, when we discovered the site, approximately 50 cm of the fluke was visible above the sediments, but upon return one week later only 20–25 cm was visible, a testament to the dynamic nature of this environment. This site provides a good case study for examining some of the historical and navigational practices associated with maritime trade at Elmina, including an investigation into its location in a likely disastrous area for ships. It is possible that there was actually a wreck and the site is simply so deeply buried that there is no additional material visible on the surface, but more work is required to answer these questions.

Another site, called the Double Anchor Site, was found in the 2007 season, but the objects were not identified as anchors until 2009. The site is comprised of two anchors with their shanks standing (relatively) vertically out of the sediments; their arms and flukes completely buried. The site is approximately 2.5 km directly south of the Elmina Castle in 12 m of water. The anchors are two different sizes and designs; the smaller of the two stands nearly 2 m out of the sediments, while the other stands 2.65 m above. They are positioned 10 m apart and oriented in the same direction, suggesting that they are associated. The basic assumption upon discovery of this site was that since the anchors were still vertical, there was likely ship structure between them that had supported them long enough to allow them to set in the sediments, even and the wood from the vessel gradually disappeared. However, there is no additional cultural material located on the surface either between or around the anchors, and a test hole between the anchors and sediment cores between and around them provided no additional cultural material (Horlings 2011:304). Clearly more research is also required concerning this site.

Located approximately 10 km west of the Elmina Castle in a shallow bay separated from the sea by a rock reef a very large anchor chain, lying in approximately 5 m water depth, was discovered. The chain measures at least 18 m long, with links measuring more than 19 cm in length, and more than 5 cm in diameter in section. One end of the chain consisted on a link that had been broken or cut and the other had a ring with a diameter of more than 30 cm. The lack of studs in links and the presence of a ring as opposed to a shackle on the end suggest that it could be of early nineteenth century construction (White 1995:21; Curryer 1999:90–106), although this does not mean that it was deposited at this location at that time. It appears to have been either snapped over the reef, or cut and discarded, and is firmly concreted to the sandstone bedrock below it in a configuration that suggests that it was pulled taut and then suddenly released. There is no indication of any other historical cultural material in the vicinity of the chain making interpretation of this site also difficult. It is difficult to imagine why a vessel large enough to handle this massive chain would have been in the vicinity of the reef, particularly on the lee side in extremely shallow and dangerously rocky water. It is possible that the remains of the vessel are still located nearby, but if that is the case, they are well hidden either under sand or among the rocks of the reef.

A final example of maritime remains in the Elmina seascape may be found in the modern wrecking of a local fishing vessel in the shore of Elmina Bay (Fig. 5.7). Local fishermen often anchor their vessels in the shallow bay outside of the Benya Lagoon as they are either preparing to go to sea or returning. During series of

Fig. 5.7   The shore wreck in Elmina (Horlings 2011:118)

powerful storms in September 2007, one of the fishing vessels broke anchor and was smashed on the rocky shore. Local tradition demands that in such an event only the owner is permitted to disturb the wreckage, and in this case there was no evidence that the owner did so. It was possible to monitor and photograph the wreck site for the 4 months and record the natural formation processes that continued to break it apart, remove loose timbers, and shift the positions of the various elements. At the end of the 2007 field season only two main elements—the bow and keel—remained on the shore. In 2009 both elements were still on the shore, although shifted in position (Horlings 2011:118). While not archaeological, this modern example illustrates the power of formation processes at work in the region, and serves to connect the sea and shore within this dynamic seascape.

The complex and intersecting processes that shaped the historical Elmina seascape include such factors as the Atlantic trade; the international relations between Africans and Europeans trading at the coast; the permanent presence of the Europeans in Elmina Castle; the resources of the Benya Lagoon; common sailing, navigational, and anchoring practices for European vessels; the central location of Elmina both in terms of the Ghanaian coast and in terms of the trade along the West African coast as a whole; and the weather and oceanographic patterns that dictated the seasons of trade. The formation processes that dictated the creation, destruction, and preservation of submerged cultural resources included all these factors; it was and is, however, the power of the sea that ultimately determined where archaeological sites are located and the conditions we find them in today. The disparate forces that affected historical sailors now affect the tangible remains of their presence across the submerged landscape. The Elmina Wreck and other submerged heritage, like the region as a whole, is the intersection of all of these processes. Understanding the complexities of all of these processes as they are presented in these sites provides a way to interrogate and transition between scales of history and events, regions and site, and micro and macro, and in the process creates a richer understanding of both the submerged archaeological record and the maritime history of the Atlantic trade in coastal Ghana.

Discussion and Conclusions: Seascape, Heritage, and Maritime Archaeology in Ghana

Full fathom five thy father lies;

Of his bones are coral made;

Those are pearls that were his eyes;

Nothing of him that doth fade,

But doth suffer a sea change

Into something rich and strange.

Shakespeare—*The Tempest*

We suspect that in his description of the dramatic transformational power of the sea Shakespeare never actually intended for the playful words of his spirit Ariel to so exquisitely depict the effects of formation processes on submerged objects, whether human or otherwise. Although acknowledging the sea's mysteriousness, his words also show an understanding of the nature of the sea even by those who were not sailors or directly connected to it, indicating the comprehension those historical sailors must have had of the underwater world above which they sailed, and of their fate should they not remain firmly in control of their ventures. It is these worlds that fledgling maritime archaeology in Ghana seeks to understand.

The historic African trade was notorious for its uncertainty, both in terms of profits and the very real possibility that those involved in the trade would never return from Africa due to mismanagement, illness, warfare, piracy, shipwreck, or any number of other mishaps (Fage 1969; Hopkins 1973:92; Williams 1994 [1944]:38; Inikori 1996:58; Eltis et al. 1999; Mancke 2004). In addition, the almost cavalier manner in which da Mota and Hair (1988:6–8) discuss competing nations intentionally sinking several different vessels on the Ghanaian coast indicates that there are undoubtedly more shipwrecks and maritime-related trade sites located there than the few sites that have been discovered to date. Added to that, the frequency with which historical vessels in general burned, were captured, sunk in storms, and suffered other mishaps, considered over a span of the nearly 400 years that Elmina was a major trading entrepôt and stop along the coast, makes it virtually impossible that there are no additional shipwreck sites located in coastal Ghana, and coastal Elmina in particular. Remote sensing data also suggest that there are significant submerged features that have yet to be identified.

What is perhaps even more intriguing than the identification of individual sites is investigating possible connections that they may have to each other, hypothetically, such as the Double Anchor Site with the Elmina Shipwreck (Horlings 2011:305). The investigations that have been conducted to date have only begun to scratch the surface of history. Future research into the events that occurred on this historically rich seascape hold the potential to offer insights into the historical maritime past that have yet to even be considered and will undoubtedly impact future interpretations of history. As Murphy (1983:69) asserts, the archaeological investigation of shipwreck and other submerged cultural resources should be "not merely the embellishment of the maritime historical record, but the elucidation of otherwise unattainable aspects of human behavior." While some insights have indeed been gained into these otherwise unattainable aspects of both human behavior and the environment in which evidence of that behavior remains, our understanding of the historic Ghanaian seascape is yet in its infancy.

Delgado (1988:3) once noted that "Not every shipwreck is old, not every old shipwreck is historic, not every shipwreck is significant." In all likelihood, the Elmina Wreck was never famous and was significant only to those who were directly affected by it. But, just as its discovery has once again placed it as an active part of Elmina's seascape, and quite apart from its intrinsic cultural and historical value, this site, along with others in the region, has been significant in demonstrating the potential for shipwreck and submerged cultural resource studies in Africa (Werz 1993:254). The application of the site formation processes framework has provided details concerning the physical remains and their tenure on the seafloor and situated those remains and the processes affecting them within the theoretical framework of larger formation processes, shedding light on the historical contexts of the past and the events that created and still affect the shipwreck site. Whether this wreck site represents the remains of *Groningen* or not is still unknown, and perhaps will never be known, but it illustrates an event in the dramatic historical eventscape of coastal Elmina.

The complex of people and events and settings that makes up the past provides the backdrop against which we can examine component parts of it represented by submerged archaeological materials. Modifying Branton's (2009:61) definition of an eventscape provides a framework for contextualizing the maritime archaeology of the Elmina seascape:

> An eventscape consists of a network of thematically connected places associated with a social group's [or several social groups'] participation in an event or related events. Eventscape encompasses not only locations with a land- or seascape but also the behaviors that took place within it, including international relations, maritime trade, and the events of daily life of those whose lives were intimately entwined with the seascape. It also encompasses the physical remains of human interactions within its sphere.

With this as a foundation, the theater of maritime history and trade can be understood as both an all-encompassing arena in which events from the past occurred, and as the repository of the remains from those events, synthesizing the interrelatedness of a number of concepts into one descriptive framework that can then be used to tell an integrated story of the past.

The integration of historical maritime practices and physical environmental processes allows for particular cultural remains to be interpreted as the remains of specific events in history, and even provides a means of conjecture for defining just what those events may have been. Although we have a relatively small sample size of unique sites and events, within the sample we have represented events that occurred in the seventeenth century (the Elmina Wreck), in the early eighteenth century (the Benya hulk), and sometime likely in the later eighteenth or early nineteenth centuries (the Single Anchor Site); the dates of the Double Anchor Site and the Chain Site are more ambiguous. The locations of each of these sites within the historical seascape speak to navigational hazards (the Chain Site and the Single Anchor Site), intentional use of the Benya Lagoon drainage (the Benya hulk), and possibly anchoring strategies, as well as the ever-present danger of sinking (the Elmina Wreck). At a more macro level, even the expanses of the submerged Elmina seascape that appear to be devoid of cultural material present us with information concerning possible activity patterns in the historical seascape related to maritime activities. Viewing the region from a bird's eye perspective suggests that while there is clearly archaeological evidence of significant activity that took place along the littoral margin, or the boundary between sea and land, the data are sparser concerning the more seaward regions.

But looking at the archaeology of the region is only one aspect of the modern seascape that now encompasses tens of thousands of people who live near and depend on it for their livelihoods. David and Thomas (2008:33) refer to significant sites and landscapes (seascapes) as "cultural resource cachements," and discuss the need for the assessment of them in terms of cultural resource management (Ash 2007:6; Jameson and Scott-Ireton 2007:2). Both Elmina Castle and neighboring Cape Coast Castle have been designated as World Heritage Sites; it is our hope that in the future the World Heritage status awarded to these monuments also includes the wider contexts in which they operated historically—the seascape and land-scapes specifically associated with international maritime trade—as part of their heritage. Underwater archaeological sites in Ghana are, for the present, exclusive in nature and protected purely because very few people there have the means and technology to explore or otherwise affect then, but this is rapidly changing, and will likely have significant impacts on underwater sites in the near future (Horlings 2012). While this is a serious concern for researchers and those interested in the past, there are far more immediate concerns for most of those who live in direct contact with the seascape today. Part of a holistic approach to this anthropological research is an awareness of the responsibility to use archaeology as a vehicle for the elucidation of the past and an opportunity to influence the concerns of the present. While goals for social and environmental responsibility, both by researcher and local, may seem lofty, they are crucial if we are to continue to invest so much in the study of the past in a place where both the past and the present have a nebulous future (Hassan 2004:319).

The presence of Elmina's 25,000 plus population living near the coast is having a significant and visible effect on submerged sites across the seascape in several ways, most obviously visible, for instance, in the extraordinary quantities of trash and debris in the water and along the coast. These items, particularly those modern items made of plastic such as bags and bottles, float easily on the currents at the surface or along the seafloor and become entrapped in submerged sites, as illus-trated in the Elmina Wreck site, as well as cause general pollution to the marine and shore environments. In addition, the (generally) unintentional modification of the submerged environment by those engaging in maritime activities, including extreme overfishing, is resulting in both the dramatic decline on marine resources, and unintentional disturbance of submerged heritage (Horlings 2011). A general lack of interest in resources other than those that provide immediately, such as fish, combined with a general lack of education on both sustainable practices and history in much of West Africa means that most people who are understandably preoccupied with the basics of life do not have the time, resources, or interest to spend on relics of the past that have no direct relation to their everyday lives. It will take a directed and very conscious approach to become effective stewards of submerged cultural history in West Africa, something that has to date occurred only minimally in discourse (Horlings 2012). As research continues to expand along the coast, including at nearby Cape Coast and Mori, this approach can and must continue to be developed and implemented.

Research on submerged cultural resource site formation necessarily intersects both cultural and natural processes (Murphy 1983, 1990; Oxley 1998a, b; Adams 2001; Gibbins and Adams 2001; Martin 2001). While it is virtually impossible to account for all factors associated with these processes individually or in combination (Ward and Larcombe 2003:1233), characterizing formation processes and their effects is an invaluable archaeological tool for understanding the dynamic and subtle processes that convert submerged remains from articulated entities into dispersed sites, and which connect the past and the present. It is indeed the intersection of ocean dynamics, dangerous seafloor and shoreline features, and shore-based trading centers that formed the maritime seascape and eventscape of the past, and which informs our understanding of it in the present.

References

Adams, John. 1966 [1823]. Remarks on the Country Extending from Cape Palmas to the River Congo with an Appendix containing an Account of the European trade with the West Coast of Africa. London: Frank Cass & Co., Ltd.

Adams, Jonathan. 2001. Ships and Boats as Archeological Source Material. *World Archaeology* 32: 292–310.

Allersma, Egge, and Wiel M.K. Tilmans. 1993. Coastal Conditions in West Africa—A Review. *Ocean and Coastal Management* 19: 199–240.

Amlolo, Daniel S. 2006. The Protection, Management and Development of the Marine and Coastal Environment of Ghana. Administering Marine Spaces: International Issues, FIG Publication No. 36. The International Federation of Surveyors (FIG), September 2006, Frederiksberg, Denmark. http://www.fig.net/pub/figpub/pub36/figpub36.htm. Accessed April 2011.

Anthony, E.J., and A.B. Blivi. 1999. Morphosedimentary Evolution of a Delta-Sourced, Drift-Aligned Sand Barrier-Lagoon Complex, Western Bight of Benin. *Marine Geology* 158: 161–176.

Anuskiewicz, Richard J. 1998. Technology, Theory, and ANALYSIS: Using Remote Sensing as a Tool for Middle-Range Theory Building in Maritime and Nautical Archaeology. In *Maritime Archaeology: A Reader of Substantive and Theoretical Contributions*, ed. Lawrence E.Babits Babits, and Hans Van Tilburg, 223–231. New York: Plenum Press.

Ash, Aiden. 2007. The Maritime Cultural Landscape of Port Willunga, South Australia. Flinders University Monograph Series no. 4. Flinders University Department of Archaeology.

Awosika, L.F., A.C. Ibe, and C.E. Ibe. 1993. Anthropogenic Activities Affecting Sediment Load Balance along the West African Coastline. In Coastlines of Western Africa, edited by Awosika, L.F., A. Chidi Ibe, and P. Schroader, pp. 26–39. New York: American Society of Civil Engineers.

Baesjou, René. 1988. The Historical Evidence in Old Maps and Charts of Africa with Special References to West Africa. History in Africa 15: 1–83.

Ballong-Wen-Mewuda, Joseph Bato'ora. 1993. São Jorge da Mina 1482–1637: La vie d'un comptoir portugais en Afrique occidentale. Fondation Calouste Gulbenkian Centre Culturel Portugais, Commission Nationale por les Commemorations de Decouvertes Portugaises, Lisbonne and Paris.

Bandinel, James. 1968 [1842]. Some Account of the Trade in Slaves from Africa: As Connected with Europe and America: From the Introduction of Trade into Modern Europe, Down to the Present Time; Especially with Reference to the Efforts made by the British Government for Its Extinction. London: Frank Cass & Co., Ltd.

Bean, Richard. 1974. A Note on the Relative Importance of Slaves and Gold in West African Exports. The Journal of African History 15: 351–356.

Blake, John W. 1977. West Africa: Quest for God and Gold, 1454–1578. London: Curzon Press.

Bold, Edward. 1823 [1819]. The Merchant's and Mariner's African Guide: Containing an Accurate Description of the Coast, Bays, Harbours, and Adjacent Islands of West Africa, with Their Corrected Longitudinal Positions, Comprising a Statement of the Seasons, Winds, and Currents, Peculiar to Each Country, to which is added A Minute Explanation of the Various Systems of Traffic, that are Adopted on the Windward and Gold Coast, as well as the Principle Ports to Leeward, Also, A Few Hints to the Mercantile Navigator, Suggesting a Means of Securing More Rapid Passages, Both To and From the Coast, than have Hitherto been Practiced. With additional observations by the Commander of a Boston vessel, and an English Gentleman, on the Coast in 1822. Salem: Cushing & Appleton.

Bourret, F.M. 1949. The Gold Coast: A survey of the Gold Coast and British Togoland. Stanford: Stanford University Press.

Bower, John. 1986. A Survey of Surveys: Aspects of Surface Archaeology in Sub-Saharan Africa. The African Archaeological Review 4: 21–40.

Branton, Nicole. 2009. Landscape Approaches in Historical Archaeology: The Archaeology of Place. In International Handbook of Historical Archaeology, ed. Teresita Majewski, and David Gaimster, 51–65. New York: Springer.

Breen, Colin, and Wes Forsythe. 2001. Management Protection of the Maritime Cultural Resource in Ireland. Coastal Management 29: 41–51.

Breen, Colin, and Paul J. Lane. 2003. Archaeological Approaches to East Africa's Changing Seascapes. World Archaeology 35: 469–489.

Brooks Jr., George E. 1970. Yankee Traders, Old Coasters and African Middlemen: A History of American Legitimate Trade with West Africa in the Nineteenth Century. Boston: Boston University Press.

Calvocoressi, David. 1968. European Traders on the Gold Coast. West African Archaeological Newsletter 10: 16–19.

Chaplin, Joyce E. 2009. The Atlantic Ocean and its Contemporary Meanings, 1492–1808. In Atlantic History: A Critical Appraisal, ed. Jack P. Greene, and Philip D. Morgan, 35–51. Oxford: Oxford University Press.

Chapman, Henry P., and Philip R. Chapman. 2005. Seascapes and Landscapes—The Siting of the Ferriby Boat Finds in the Context of Prehistoric Pilotage. International Journal of Nautical Archaeology 34: 43–50.

Cook, Gregory D. 2012. West Africa and the Atlantic World: Maritime Archaeological Investigations at Elmina, Ghana. Ph.D. dissertation, Department of Anthropology, Syracuse University.

Cook, Gregory D., and Sam Spiers. 2004. Central Region Project: Ongoing Research on Early Contact, Trade and Politics in Coastal Ghana, AD 500–2000. Nyame Akuma 61: 17–28.

Coombs, Douglas. 1963. *The Gold Coast, Britain and the Netherlands, 1850–1874*. London: Oxford University Press.

Cooney, Gabriel. 2003. Introduction: Seeing Land from the Sea. *World Archaeology* 35: 323–328.

Cotsell, George N.A. 1858. *A Treatise on Ships' Anchors. With Numerous Illustrations*. London: John Weale.

Crooks, John Joseph. 1973 [1923]. Records Relating to the Gold Coast Settlements from 1750 to 1874. London: Frank Cass.

Crumley, Carole L., and William H. Marquardt. 1990. Landscape: A Unifying Concept in Regional Analysis. In Interpreting Space: GIS and Archaeology, edited by Kathleen M.S. Allen, Stanton W. Gree, and Ezra B.W. Zubrow, pp. 73–79. London: Taylor & Francis.

Cunliffe, Barry. 2001. *Facing the Ocean: The Atlantic and its Peoples 8000 BC–AD 1500*. Oxford: Oxford University Press.

Curryer, Betty Nelson. 1999. *Anchors: An Illustrated History*. Annapolis, Maryland: Naval Institute Press.

Curtin, Philip D. 1984. *Cross-Cultural Trade in World History*. Cambridge: Cambridge University Press.

da Mota, A. T., & Hair, P. E. (1988). *East of Mina: Afro-European relations on the Gold Coast in the 1550 s and 1560 s. African Studies Program*. Madison: University of Wisconsin.

David, Bruno, and Julian Thomas. 2008. Landscape Archaeology: An Introduction. In Handbook of Landscape Archaeology, edited by Bruno David and Julian Thomas, pp. 27–43. Walnut Creek, CA: Left Coast Press.

de Marees, Pieter. 1987. Description and Historical Account of the Gold Kingdom of Guinea (1602). Translated from the Dutch and edited by Albert van Dantzig and Adam Jones. Oxford: Oxford University Press.

DeCorse, Christopher R. 2001. An Archaeology of Elmina: Africans and Europeans on the Gold Coast. 1400–1900. Washington: Smithsonian Institute Press.

DeCorse, Christopher R. 2010. Early Trade Posts and Forts of West Africa. In *First Forts: Essays on the Archaeology of Proto-Colonial Fortifications*, ed. Eric Klingelhofer, 209–233. Brill: Leiden.

DeCorse, Christopher R., Edward Carr, Gerard Chouin, Gregory Cook, and Sam Spiers. 2000. Central Region Project, Coastal Ghana. *Nyame Akuma* 53: 6–11.

DeCorse, Christopher R., Gregory Cook, Rachel Horlings Andrew Pietruszka, and Samuel Spiers. 2009. Transformation in the Era of the Atlantic World: The Central Region Project, Coastal Ghana 2007–2008. *Nyame Akuma* 72: 85–93.

Delgado, James P. 1988. The Value of Shipwrecks. In Historical shipwrecks: Issues in Management, edited by Joy Waldron Murphy, pp.1–10. Washington, DC: Partners for Livable Places and the National Trust for Historic Preservation Maritime Department.

Delgado, James P., and Mark Staniforth. 2002. Underwater Archaeology. In The Encyclopedia of Life Support Systems (online). UNESCO, Paris. http://www.eolss.co.uk. Accessed Jan 2011.

Dellino, Virginia, and María Luz Endere. 2001. The HMS Swift Shipwreck: The Development of Underwater Heritage Protection in Argentina. *Conservation and Management of Archaeological Sites* 4: 219–231.

Dellino-Musgrave, Virginia E. 2006. *Maritime Archaeology and Social Relations: British Action in the Southern Hemisphere*. New York: Springer.

den Heijer, Henk. 2003. The West African Trade of the Dutch West India Company, 1674–1740. In 2003 Riches from Atlantic Commerce: Dutch Transatlantic Trade and Shipping, 1585-1817, edited by Johannes Postma and Victor Enthoven, pp. 139–169. Brill, Leiden.

Dickson, K.B. 1965. Evolution of Seaports in Ghana: 1800–1928. *Annals of the Association of American Geographers* 55: 98–111.

Dincauze, D.F. 2000. *Environmental Archaeology: Principles and Practice*. Cambridge: Cambridge University Press.

Ellis, Alfred Burdon. 1969 [1893]. A History of the Gold Coast of West Africa. New York: Negro Universities Press.

Eltis, David, Stephen D. Behrendt, David Richardson, and Herbert S. Klein. 1999. The Atlantic Slave Trade: A Database on CD-ROM. Cambridge: Cambridge University Press.

Fage, J.D. 1969. *A History of West Africa: An Introductory Survey*, 4th ed. Cambridge: Cambridge University Press.

Feinberg, Harvey M. 1989. Africans and Europeans in West Africa: Elminans and Dutchmen on the Gold Coast during the Eighteenth Century. *Transactions of the American Philosophical Society* 79.

Firth, Anthony. 1995. Three Facets of Maritime Archaeology: Society, Landscape and Critique. Department of Archaeology, Southampton: Southampton University. http://avebury.arch.soton. ac.uk/Research/Firth. Accessed January 7, 2008.

Flatman, Joe. 2003. Cultural Biographies, Cognitive Landscapes and Dirty Old Bits of Boat: 'Theory' in Maritime Archaeology. *The International Journal of Nautical Archaeology* 32: 143–157.

Flatman, Joe, Mark Staniforth, David Nutley, and Debra Shefi. 2005. Submerged Cultural Landscapes. Humanities Research Centre for Cultural Heritage and Cultural Exchange 'Understanding Cultural Landscapes' Symposium Report. http://wwwehlt.flinders.edu.au/humanities/exchange/asri/ucl_symp_pdf/2005_UCL_MS.pdf. Accessed May, 2009.

Foxhall, Lin. 2000. The Running Sands of Time: Archaeology and the Short-Term. *World Archaeology* 31: 484–498.

Garrison, Ervan G. 1998. A Diachronic Study of Some Historical and Natural Factors Linked to Shipwreck Patterns in the Northern Gulf of Mexico. In *Maritime Archaeology: A Reader of Substantive and Theoretical Contributions*, ed. Lawrence E. Babits, and Hans Van Tilburg, 303–321. New York: Plenum Press.

Gibbins, David, and Jonathan Adams. 2001. Shipwrecks and maritime archaeology. *World Archaeology* 32: 279–291.

Gilchrist, Roberta. 2005. Introduction: Scales and Voices in World Historical Archaeology. *World Archaeology* 37: 329–336.

Gladfelter, Bruce G. 1977. Geoarchaeology: The Geomorphologist and Archaeology. *American Antiquity* 42: 519–538.

Goldberg, Paul, David T. Nash and Michael D. Petraglia (1993) Preface. In Formation Processes in Archaeological Context. Monographs in World Archaeology No. 17, edited by Paul Goldberg, David T. Nash and Michael D. Petraglia, pp. vii–ix. Madison: Prehistory Press.

Gosden, Chris, and Lesley Head. 1994. Landscape—A usefully ambiguous concept. *Archaeology in Oceania* 29: 113–116.

Gosden, Chris, and Karola Kirsanow. 2006. Timescales. In Confronting Scale in Archaeology: Issues of Theory and Practice, edited by Gary Lock and Brian Leigh Molyneaux, pp. 27–37.

Gould, Richard A. (2001). From Sail to Steam at Sea in the Late Nineteenth Century. In Anthropological Perspectives on Technology, edited by Michael Brian Schiffer, pp. 193–213. Albuquerque: University of New Mexico Press.

Green, Jeremy. 2004. *Maritime Archaeology: A Technical Handbook*, 2nd ed. London: Academic Press.

Gu, Gu, and Robert F. Adler. 2004. Seasonal Evolution and Variability Associated with the West African Monsoon System. *Journal of Climate* 17: 3364–3377.

Gutkind, Peter C. W. 1989. The Canoemen of the Gold Coast (Ghana): A Survey and an Exploration in Precolonial AfricanLabour History (Les piroguiers de la Côte de l'Or (Ghana): enquête et recherché d'histoire dutravail en Afrique précoloniale). Cahiers d'Études Africaines 29, Cahier 115/116:339–376.

Hair, Paul E. 1994. *The Founding of the Castelo de São Jorge da Mina: An Analysis of the Sources*. African Studies Program: University of Wisconsin, Madison.

Hair, Paul E.H., Adam Jones, and Robin Law (eds.). 1992. *Barbot on Guinea: The Writings of Jean Barbot on West Africa 1678–1712*. London: The Hakluyt Society.

Harris, Trevor M. (2006). Scale as Artifact: GIS, Ecological Fallacy, and Archaeological Analysis. In Confronting Scale in Archaeology: Issues of Theory and Practice, edited by Gary Lock and Brian Leigh Molyneaux, pp. 39–53. New York: Springer.

Hassan, F.A. 2004. Ecology in Archaeology: From Cognition to Action. In *A Companion to Archaeology*, ed. John Bintliff, 311–333. Malden, Massachusetts: Blackwell Publishing.

Head, Lesley. 2008. Geographical Scale in Understanding Human Landscapes. In *Handbook of Landscape Archaeology*, ed. Bruno David, and Julian Thomas, 379–385. Walnut Creek, CA: Left Coast Press.

Hood, J. Edward. 1996. Social relations and the Cultural Landscape. In Landscape Archaeology: Reading and Interpreting the American Historical Landscape, edited by Rebecca Yamin and Karen Bescherer Metheny, pp. 121–146. Knoxville: The University of Tennessee Press.

Hopkins, Anthony G. 1973. *An Economic History of WEST Africa*. New York: Columbia University Press.

Horlings, Rachel L. 2009. Technical Brief: An Effective Diver-Operated Coring Device for Underwater Archaeology. *Technical Briefs in Historical Archaeology* 4: 1–6.

Horlings, Rachel L. 2011. Of His Bones are Coral Made: Submerged Cultural Resources, Site Formation Processes, and Multiple Scales of Interpretation in Coastal Ghana. PhD dissertation. New York: Department of Anthropology, Syracuse University, Syracuse.

Horlings, Rachel L. 2012. Maritime Cultural Resource Investigation, Management, and Mitigation in Coastal Ghana. *Journal of Maritime Archaeology* 7: 141–164. doi:10.1007/s11457-012-9086-9.

Horlings, Rachel L. 2013. Archaeological Micro-Sampling by Means of Sediment Coring at Submerged Sites. *Geoarchaeology* 28: 308–315.

Horlings, Rachel L., Darren Kipping, Casper Toftgaard Neilsen, and Kira Kaufmann. 2011. Missing Shipwrecks, Methods or Imagining? A Preliminary Report on Maritime Archaeological Surveys in Coastal Ghana, 2009. *Nyame Akuma* 75: 2–10.

Inikori, Joseph E. 1996. Measuring the Unmeasured Hazards of the Atlantic Slave Trade: Documents Relating to the British Trade. *Revue française d'histoire d'outre-mer* 312: 53–92.

John, David M., Diana Lieberman, and Milton Lieberman. 1977. A Quantitative Study of the Structure and Dynamics of Benthic Subtidal Algal Vegetation in Ghana (Tropical West Africa). *The Journal of Ecology* 65: 497–521.

Johnson, Matthew. 2007. *Ideas of Landscape*. Oxford: Blackwell Publishing.

Kelly, Kenneth G. 2004. The African Diaspora Starts Here: Historical Archaeology of Coastal West Africa. In African Historical Archaeologies, edited by Andrew M. Reid and Paul J. Lane, pp. 219–241. New York: Kluwer Academic/Plenum Publishers.

Kelly, Kenneth G., and Norman, Neil L. 2007. Historical Archaeology of Landscape in Atlantic Africa. In Envisioning Landscape: Situations and Standpoints in Archaeology and Heritage, edited by Dan Hicks, Laura McAtackney, Graham Fairclough. Walnut Creek, CA: Left Coast Press.

Knapp, A. Bernard, and Wendy Ashmore. 1999. Archaeological Landscapes: Constructed, Conceptualized, Ideational. In Archaeologies of Landscape: Contemporary Perspectives, edited by Wendy Ashmore and A. Bernard Knapp, pp. 1–30. Oxford: Blackwell Publishers Ltd.

Koranteng, Kwame A. 2001. Structure and Dynamics of Demersal Assemblages on the Continental Shelf and Upper Slope off Ghana, West Africa. *Marine Ecology Progress Series* 220: 1–12.

Lawrence, A.W. 1963. *Trade Castles & Forts of West Africa*. California: Stanford University Press.

Lenihan, Daniel J., and Larry Murphy. 1998. Considerations for Research Designs in Shipwreck Archaeology. In Maritime Archaeology: A Reader of Substantive and Theoretical Contributions, edited by Lawrence E. Babits and Hans Van Tilburg, pp. 233–239. New York: Plenum Press.

Mancke, Elizabeth. 2004. Oceanic Space and the Creation of a Global International System, 1450–1800. In *Maritime History as World History*, ed. Daniel Finamore, 149–166. Gainesville: University of Florida.

Marquardt, William H., and Carole L. Crumley. 1987. Theoretical Issues in the Analysis of Spatial Patterning. In Regional Dynamics: Burgundian Landscapes in Historical Perspective, edited by Carole L. Crumley and William H. Marquardt, pp. 1–18. San Diego: Academic Press, Inc.

Martin, R. Montgomery 1837. History of the British Possessions in the Indian & Atlantic Oceans; Comprising Ceylon, Penang, Malacca, Sincapore [sic], The Falkland Islands, St. Helena, Ascension, Sierra Leone, The Gambia, Cape Coast Castle, &c. &c. London: Whitaker & Co..

Martin, Colin J.M. 2001. De-Particularizing the Particular: Approaches to the Investigation of Well-Documented Post-Medieval Shipwrecks. *World Archaeology* 32: 383–399.

McCarthy, Mike. 2006. Maritime Archaeology in Australasia: Reviews and Overviews. In *Maritime Archaeology: Australian Approaches*, ed. Mark Staniforth, and Michael Nash, 1–11. New York: Springer.

McCarthy, Mike. 2008. Boundaries and the Archaeology of Frontier Zones. In *Handbook of Landscape Archaeology*, ed. Bruno David, and Julian Thomas, 202–209. Walnut Creek, CA: Left Coast Press.

McConkey, R., and T. McErlean. 2007. Mombasa Island: A Maritime Perspective. *International Journal of Historical Archaeology* 11: 99–121.

McErlean, T., R. McConkey, and W. Forsythe. 2003. *Strangford Lough: An Archaeological Survey of the Maritime Cultural Landscape*. Belfast: The Blackstaff Press.

McGhee, Fred L. 2007. Maritime Archaeology and the African Diaspora. In *Archaeology of Atlantic Africa and the African Diaspora*, ed. Akinwumi Ogundiran, and Toyin Falola, 384–394. Bloomington: Indiana University Press.

Meredith, Henry. 1967 [1812]. An Account of the Gold Coast of Africa: With a Brief History of the African Company. London: Longman, Hurst, Rees, Orme, and Brown, Paternoster Row.

Mitchell, Peter. 2005. *African Connections: Archaeological Perspectives on Africa and the Wider World*. Walnut Creek: AltaMira Press.

Morgan, Philip D. 2009. Africa and the Atlantic, c. 1450–1820. In Atlantic History: A Critical Appraisal, edited by Jack P. Greene and Philip D. Morgan, pp. 223–248. Oxford: Oxford University Press.

Morgan, Philip D., and Jack P. Greene. 2009. Introduction: The Present State of the Atlantic History. In *Atlantic History: A Critical Appraisal*, ed. Jack P. Greene, and Philip D. Morgan, 3–33. Oxford: Oxford University Press.

Murphy, Larry E. 1983. Shipwrecks as Data Base for Human Behavioral Studies. In *Shipwreck Anthropology*, ed. Richard A. Gould, 65–89. Albuquerque: University of New Mexico Press.

Murphy, Larry E. 1990. 8SL17: Natural Site-Formation Processes of a Multiple-Component Underwater Site in Florida (Southwest Cultural Resources Center Professional Papers, No. 39). Santa Fe, New Mexico.

Nathan, Matthew. 1904. Dutch and English on the Gold Coast in the Eighteenth Century. *Journal of the Royal African Society* 3: 325–351.

Nørregård, Georg. 1966. *Danish Settlements in West Africa 1658–1850*. Translated by Sigurd Mammen: Boston University Press, Boston.

O'Shea, John M. 2002. The Archaeology of Scattered Wreck-Sites: Formation Processes and Shallow Water Archaeology in Western Lake Huron. *International Journal of Nautical Archaeology* 31: 211–227.

Ogundiran, Akinwumi, and Toyin Falola. 2007. Pathways in the Archaeology of Transatlantic Africa. In *Archaeology of Atlantic Africa and the African Diaspora*, ed. Akinwumi Ogundiran, and Toyin Falola, 3–45. Bloomington: Indiana University Press.

Opoku-Ankomah, Yaw, and Ian Cordery Cordery. 1994. Atlantic Sea Surface Temperatures and Rainfall Variability in Ghana. *Journal o Climate* 7: 551–558.

Oxley, Ian. 1998a. The Environment of Historical Shipwreck sites: A Review of the Presentation of Materials, Site Formation and Site Environmental Assessment (Unpublished Master's thesis). Fife, Scotland: School of Geography and Geosciences, University of St Andrews.

Oxley, Ian. 1998b. The Investigation of Factors that Affect the Preservation of Underwater Archaeological Sites. In Maritime Archaeology: A Reader of Substantive and Theoretical Contributions, edited by Lawrence E. Babits and Hans Van Tilburg, pp. 523–529. New York: Plenum Press.

Parker, A.J. 2001. *Maritime Landscapes. Landscapes* 1: 22–41.

Pearson, Patricia. 2008. The World of the Atlantic before the "Atlantic World": Africa, Europe, and the Americas before 1450. In *The Atlantic World 1450–2000*, ed. Toyin Falola, and Kevin D. Roberts, 3–26. Bloomington: Indiana University Press.

Pering, Richard Esq. 1819. A Treatise on the Anchor, Shewing How the Component Parts should be Combined to Obtain the Greatest Power, and Most Perfect Holding, with A Table of Dimensions, Graduated from 1 to 95 Cwt. and A Schedule of Proportionate Weights of Anchors Suitable to the Tonnage of Every Class of Vessel, Both as to His Majesty's and Merchant's Ships, with A Recommendation for their Better Security in Anchoring. Also, Some Observations on the Chain Cable, Shewing the Proportion it Bears to that of Hemp, and to the Weight of the Anchor. Congdon and Hearle, London.

Pietruszka, Andrew. 2011. Artifacts of Exchange: A Multi-Scalar Approach to Maritime Archaeology at Elmina, Ghana (Ph.D. dissertation). Department of Anthropology, Syracuse University.

Posnansky, Merrick, and Christopher DeCorse. 1986. Historical Archaeology in Sub-Saharan Africa: A Review. *Historical Archaeology* 20(1): 1–14.

Preucel, Robert W., and Lynn Meskell. 2009. Places. In A Companion to Social Archaeology, edited by Lynn Meskell and Robert W. Preucel, pp. 215–229.

Priestley, Margaret. 1969. *West African Trade and Coast Society: A Family Study*. London: Oxford University Press.

Quine, Timothy A. 1995. Soils Analysis and Archaeological Site Formation Studies. In Archaeological Sediments and Soils: Analysis, Interpretation and Management, edited by Anthony J. Barham and Richard I. McPhail, pp. 77–98. Papers from the Tenth Anniversary Conference of the Association for Environmental Archaeology held at the Institute of Archaeology, UCL, July, 1989. Institute of Archaeology, University College, London.

Rapp Jr., George R. 2000. Geoarchaeology. In *Archaeological Method and Theory: An Encyclopedia*, ed. L. Ellis, 237–244. New York: Garland Publishing Inc.

Rawley, James A., and Stephen D. Behrendt. 2005. *The Atlantic Slave Trade: A History*, Revised ed. Lincoln: University of Nebraska Press.

Reitz, Elizabeth J., Lee A. Newsom, and Sylvia Scudder. 1996. Issues in Environmental Archaeology. In *Case Studies in Environmental Archaeology*, ed. Elizabeth J. Reitz, Lee A. Newsom, and Sylvia Scudder, 3–16. New York: Plenum Press.

Richard, François Gilles. 2006. From Cosaan to Colony: Exploring Archaeological Landscape Formation and Socio-Political Complexity in the Siin (Senegal), AD 500–1900 (Unpublished dissertation). Syracuse University.

Rømer, Ludewig Ferdinand. 2000. *A reliable Account of the Coast of Guinea (1760), Translated from the Danish and edited by Selena Axelrod Winsnes*. Oxford: Oxford University Press.

Rubin, Norman. 1971. The Anchor. *Nautical Research Journal* 18: 230–250.

Russell-Wood, A.J.R. 2009. The Portuguese Atlantic, 1415–1808. In *Atlantic History: A Critical Appraisal*, ed. J.P. Greene, and P.D. Morgan, 81–109. Oxford: Oxford University Press.

Schiffer, M.B. 1987. *Formation Processes of the Archaeological Record*. Salt Lake City: University of Utah Press.

Schiffer, Michael B. 1996. Formation Processes of the Historical and Archaeological Record. In Learning from Things: Method and Theory of Material Culture Studies, edited by W. David Kingery, pp.73–80. Washington: Smithsonian Institute Press.

Schwartz, Stuart B., and Johannes Postma. 2003. The Dutch Republic and Brazil as Commercial Partners on the West African Coast during the Eighteenth Century. In *2003 Riches from Atlantic Commerce: Dutch Transatlantic Trade and Shipping, 1585–1817*, ed. Johannes Postma, and Victor Enthoven, 171–199. Brill: Leiden.

Jameson, John H. Jr., and Della A. Scott-Ireton. 2007. Introduction: Imparting Values/Making Connections. In Out of the Blue: Public Interpretation of Maritime Cultural Resources, edited by John H. Jameson Jr. and Della A. Scott-Ireton, pp. 1–6. New York: Springer.

Semple, Ellen Churchill. 1931. *The Geography of the Mediterranean Region: Its Relation to Ancient History*. New York: H. Holt and Co.

Smith, Robert. 1970. The Canoe in West African History. *The Journal of African History* 11: 515–533.

Stahl, Ann Brower. 2001. *Making History in Banda: Anthropological Visions of Africa's Past*. New York: Cambridge University Press.

Staniforth, Mark. 2003. *Material Culture and Consumer Society: Dependent Colonies in Colonial Australia*. New York: Kluwer Academic/Plenum Publishers.

Staski, Edward. 2000. Archaeological Sites, Formation Processes. In *Archaeological Method and Theory: An Encyclopedia*, ed. Linda Ellis, 39–44. New York: Garland Publishing Inc.

Stein, Julie K. 1987. Deposits for Archaeologists. *Advances in Archaeological Method and Theory* 11: 337–395.

Stein, Julie K., and Willaim R. Farrand. 1985. Context and Geoarchaeology: An Introduction. In Archaeological Sediments in Context, edited by Julie K. Stein and William R. Farrand, pp.1–3. Peopling of the Americas edited volume series: Vol1. University of Maine, Orono.

Steinberg, Philip E. 2001. *The Social Construction of the Ocean*. Cambridge: Cambridge University Press.

Stewart, David J. 1999. Formation Processes Affecting Submerged Archaeological Sites: An overview. Geoarchaeology: An International Journal 14: 565–587.

Thomas, Charles W. (1969 [1860]) Adventures and Observations on the West Coast of Africa, and Its Islands. Historical and Descriptive Sketches of Madiera, Canary, Biafra and Cape Verd Islands; Their Climates, Inhabitants and productions. Accounts of Places, Peoples, Customs, Trade, Missionary Operations, Etc., Etc. On that Part of the African Coast Lying Between Tangier, Morocco and Benguela. New York: Negro Universities Press.

Thomson, Anthony S. 1902. Anchors: Old Forms and Recent Developments. *Journal of the Royal United Service Institution* 46: 1400–1418.

Thornton, John. 1992. *Africa and Africans in the Making of the Atlantic World, 1400–1680*. Cambridge: Cambridge University Press.

Torrence, Robin. 2002. Cultural Landscapes on Garua Island, Papua New Guinea. *Antiquity* 76: 766–776.

Unger, Richard W. 1998. *Ships and Shipping in the North Sea and Atlantic, 1400–1800*. Sydney: Ashgate Variorum.

US Navy Hydrographic Office. 1951. *Sailing Directions for the SOUTHWEST Coast of Africa: Cape Palmas to the Cape of Good Hope*. Washington: United States Government Printing Office.

van Dantzig, Albert. 1980. *Forts and Castles of Ghana*. Accra: Sedco Publishing Limited.

Van de Noort, Robert. 2004. An Ancient Seascape: The Social Context of Seafaring in the Early Bronze Age. *World Archaeology* 35: 404–415.

Van Den, Boogaart. 1992. The Trade Between Western Africa and the Atlantic World, 1600–90: Estimates of Trends in Composition and Value. *The Journal of African History* 33: 369–385.

van Dommelen, Peter. 1999. Exploring Everyday Places and Cosmologies. In Archaeologies of Landscape: Contemporary Perspectives, edited by Wendy Ashmore and A. Bernard Knapp, pp. 277–285. Oxford: Blackwell Publishers Ltd.

Veth, Peter. 2006. Theoretical Approaches. In *Maritime Archaeology: Australian Approaches*, ed. Mark Staniforth, and Michael Nash, 13–26. New York: Springer.

Vogt, John. 1979. *Portuguese Rule on the Gold Coast, 1469–1682*. Athens: University of Georgia Press.

Wandsnider, LuAnn. 1998. Regional Scale Processes and Archaeological Landscape Units. In Unit Issues in Archaeology: Measuring Time, Space, and Material, edited by Ann. F. Ramenofsky and Anastasia Steffan, pp. 87–102. Salt Lake City: University of Utah Press.

Ward, Jason. 2003. The Other Atlantic World. History. *Compass* 1: 1–6.

Ward, Ingrid, and Piers Larcombe. 2003. A Process-Oriented Approach to Archaeological Site Formation: Application to Semi- arid Northern Australia. *Journal of Archaeological Science* 30: 1223–1236.

Weggel, J. Richard. 1972. An Introduction to Oceanic Water Motions and their Relation to Sediment Transport. In Shelf Sediment Transport: Process and Pattern, edited by J. P. Swift, David B. Duane and Orrin H. Pilkey, pp. 1–20. Strousburg: Dowden, Hutchinson & Ross, Inc.

Werz, Bruno E.J.S. 1993. Shipwrecks of Robben Island, South Africa: An Exercise in Cultural Resource Management in the Underwater Environment. *The International Journal of Nautical Archaeology* 22: 245–256.

Westerdahl, Christer. 1992. Maritime Cultural Landscape. *International Journal of Nautical Archaeology* 21: 5–14.

White, K.L. 1989. Geomorphic Processes and Archaeological Site Preservation. In Interdisciplinary Workshop on the Physical-Chemical-Biological Processes Affecting Archaeological Sites, Report edited by Christopher C. Matthewson, pp. 159–167. US Army Corps of Engineers.

White, David. 1995. Anchor Work in the Royal Navy. *Model Shipwright* 92: 21–25.

Williams, Eric. 1994 [1944]. Capitalism & Slavery, 2nd ed. Chapel Hill: The University of North Carolina Press.

Yarak, Larry W. 2003. A West African Cosmopolis: Elmina (Ghana) in the Nineteenth Century. Paper presented at Seascapes, Littoral Cultures, and Trans-Oceanic Exchanges, Library of Congress, Washington D.C., February 12–15, 2003.

Zook, George Frederick. 1919. On the West Coast of Africa. *The Journal of Negro History* 4: 163–205.

Chapter 6
Environment and Agency in the Formation of the Eastern Ship Graveyard in the Central Basin at Thonis-Heracleion, Egypt

Damian Robinson, Franck Goddio and David Fabre

Introduction

Beneath the waters of Aboukir Bay at the edge of Egypt's north-western Nile Delta, lies a vast submerged landscape. Here in this marshy and lagoonal area a major port developed in the early first millennium BC at the end of the Canopic branch of the Nile. With its close links to the center of regional and then Egyptian political power in Saïs, Thonis-Heracleion provided access to the increasingly important Greek trading networks, surveillance over vessels entering or leaving the Nile, and could also be relied upon to provide a defensive bastion against hostile forces—both real and spiritual (Goddio 2015; Goddio and Fabre 2015). Although an ideal location for a port, Thonis-Heracleion developed through the decisions and actions of its inhabitants and their political leaders (Goddio et al. 2015). Consequently, it is this interplay between environment and agency in the formation of the maritime landscape of this city and in its excavation and interpretation that is at the heart of this chapter. It moves from wide scale landscape formation processes that address the issue of why the city was located at the edge of the Mediterranean and why the landscape collapsed into it. It will then investigate what this means for the formation of the archaeological record. The chapter then turns to the extensive nautical archaeological record from the port city and again considers similar issues: what were the environmental factors coupled with human decision-making that resulted in the findings of over 700 ancient anchors and 69 ships in the waterways of Thonis-Heracleion? Finally, the level of analysis will be focused even more tightly

D. Robinson (✉)
Oxford Centre for Maritime Archaeology, Institute of Archaeology,
University of Oxford, 36 Beaumont St, Oxford OXI 2PG, UK
e-mail: Damian.robinson@arch.ox.ac.uk

F. Goddio · D. Fabre
European Institute for Underwater Archaeology, 17 rue Thouin,
75005 Paris, France

© Springer International Publishing AG 2017
A. Caporaso (ed.), *Formation Processes of Maritime Archaeological Landscapes*,
When the Land Meets the Sea, DOI 10.1007/978-3-319-48787-8_6

onto a single element within the landscape, Ship 43, which is part of the eastern ship graveyard in the Central Basin of the port. This vessel is currently under excavation by the University of Oxford's Centre for Maritime Archaeology (OCMA), with one of the core aims of this work being the understanding of the relationship of this vessel to the rest of the graveyard and interpreting why so many ships of the same type and age were sunk in this particular location in the landscape.

Survey and Excavation at Thonis-Heracleion

Research into the submerged landscape beneath the waters of Aboukir Bay (Fig. 6.1) by the European Institute for Underwater Archaeology (IEASM), under the overall direction of Franck Goddio, in cooperation with the Supreme Council of Antiquities of Egypt's Ministry of Antiquities (SCA), began in 1996 and continues to this day. Building on a research methodology developed during investigations into the similarly submerged landscape of Alexandria's *Portus Magnus*, the great eastern port (Goddio 1998; Fabre and Goddio 2010), the IEASM conducted an extensive geophysical survey throughout the entire 110 km^2 survey area at the western end of the bay. The results of this were used to guide more detailed

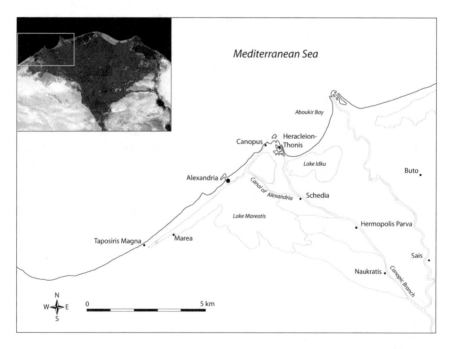

Fig. 6.1 The North-Western Nile Delta. After Mckenzie 2007, Fig. 35 with Earth Sat image of the Nile Delta (© EarthSat Natural Vue)

campaigns of geophysical, geomorphological and diver-led survey, as well as targeted stratigraphic excavation (Goddio 2007:8–17). The initial aims of the work to were twofold: to map the subsurface topography and identify areas of potential archaeological remains across the landscape that required further investigation; and to come to an understanding of the formation processes that resulted in the patterns of occupation and the reasons behind their eventual submergence. The results of this research revealed the complex interplay between environmental and anthropogenic factors in which human agency played a key role in the formation of this landscape.

The size of the survey area, in combination with the limited visibility in the shallow waters of Aboukir Bay, which is often less than a meter, and the high sediment outflow from the Nile that has covered the entire archaeological landscape in sands and alluvium, largely precluded the use of visual survey techniques (Green 2004; Bowens 2008). Consequently, investigations were guided by the geophysical survey and promising-looking anomalies investigated, of which the largest group were located 6.5–7 km east of the modern coastline. In this location large-scale magnetic anomalies suggested the presence of a settlement on a group of islands between two depressions, the easternmost of which appeared to be connected to the paleo-riverbed of the Canopic branch of the Nile via channels to the northeast and southeast (Fig. 6.2). A series of geological boreholes were taken across this area to

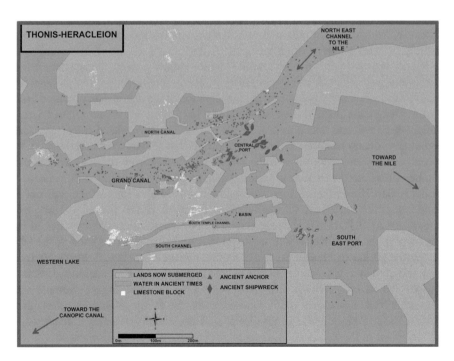

Fig. 6.2 The landscape of Thonis-Heracleion showing the locations of the land and the nautical assemblages (*map* F. Goddio © F. Goddio/IEASM)

characterize the subsurface sediments, the results of which supported the interpretations pertaining to the overall shape of the port landscape, as well as suggesting reasons for its submergence (Stanley 2007). Work continues at Thonis-Heracleion and in yearly missions the ancient topography of the city and its harbor are refined still further. The survey is revealing a landscape that is an archaeological palimpsest whose development can be separated into a number of distinct temporal phases (Goddio 2015; Robinson and Goddio 2015:11).

Excavation on the Central Island revealed extensive architectural remains of a large temple complex with its temenos wall and associated buildings. From within this area two objects were recovered that provided the names of the settlement under investigation, as well as its location at the 'edge of the Sea of the Greeks'. The first of these was a naos, a small shrine in which a statue of the major deity of the temple was housed, which was dedicated to the god Amun-Gereb. The previously discovered Decree of Canopus (discovered on land in 1881 by Maspero) reveals that the temple to Amun-Gereb was located in a settlement called 'Heracleion' (Yoyotte 2004; Goddio 2008:44–46). This interpretation was further nuanced by the discovery of a stele during excavations in the temple of Amun-Gereb recording a decree of pharaoh Nectanebo I, which states that it was set up in a settlement called Thonis (Yoyotte 2008a:236–240); von Bomhard 2012, 2015). Both names relate to the same place, with Heracleion being its Greek name and Thonis its Egyptian, which neatly describes the situation in the north-western Delta during the Late Period when it was becoming increasingly opened to Greek influence, which culminated in the Ptolemaic era when Egypt was ruled by a dynasty with Macedonian origins.

The datable archaeological evidence from the excavations in Thonis-Heracleion also firmly places its phases of occupation within the period of the first millennium BC. As the location of a major port, a consideration of the nautical assemblage from Thonis-Heracleion is instructive, as the ships that have been found within its harbors, waterways, and anchorages have all been radiocarbon dated to between the eighth and second centuries BC. The majority of the vessels are from the sixth to second century BC, with 40% of the assemblage dating from the Late Period and another 40% from the Ptolemaic period (Fabre and Belov 2011:108–109; Fabre and Goddio 2013:70). Evidence from the ceramics would also suggest that the major phase of settlement spans the period from the sixth to the second century BC (Grataloup 2010, 2015). This interpretation is corroborated and further enhanced by the numismatic evidence, which indicates that coins begin to be used at the site during the period of Persian domination and continued into the Ptolemaic period, after which there is a dramatic decline in coin loss (and therefore use) in the second half of the second century BC, with only a small scattering of Roman and Byzantine coins (Meadows 2015). Consequently, the dating evidence suggests that the site rises to prominence in the Egyptian Late Period, becoming a major economic center in the Persian period, and substantially declined in importance in the Ptolemaic to the point that there was only a scattered community inhabiting the once great seaport in the Byzantine period. It is important to note, however, that only a small proportion of the overall site—perhaps only one per cent—has been systematically investigated and it is possible that there still could be an earlier settlement and associated harbor, awaiting discovery in some yet to be examined area.

The initial archaeological work at the site was also accompanied by a systematic geological survey that aimed at understanding the processes that resulted in the submergence of the landscape of the Canopic region. That this was, in part, gradual is seen in the area of the north wall of the temple of Amun-Gereb where the archaeological layers are covered by marine sediments and mussels (Goddio 2007:83). This would suggest that sea level in the area gradually rose over time until the area north of the Grand Canal became, at first, partially submerged and part of the intertidal zone where mussels could grow. Indeed, Sophronios of Jerusalem described a similar situation at the nearby city of Canopus where a temple was still standing at the shoreline at the beginning of the seventh century BC (Bernand 1970:215). The gradual rise in sea level can be accounted for by the melting of the polar ice caps and the thermal expansion of ocean waters, coupled with the contemporaneous lowering of the land at the Nile Delta margin due to the compaction of the water laden sediments and the isostatic lowering of the landmass (Stanley 2007:46). The archaeological remains at Thonis-Heracleion are found at depths of between 5 and 7 m present sea level and if we assume that the temple was initially built at a height of 2 m above sea level, this would suggest a total lowering of the landscape by about 8 m. Yet, taking into account the rate of annual sea level rise for the region it has been calculated that this would only result in a total rise of about 4.6 m, suggesting that there is another element that needs to be accounted for. The second, and far more dramatic, factor in the submergence of the landscape is the sudden failure of the poorly consolidated deltaic sediments resulting in their vertical displacement downward by as much as 3 m or more. This is a natural phenomenon that can occur from the weight of the sediments alone, but which can also be triggered by earthquake tremors, the sudden overloading of the sediments by storm surges, tsunamis or heavy floods, or indeed the construction of heavy buildings (Stanley 2007:51–57).

Landscape Formation Processes

The Structural Remains from the City

The catastrophic collapse of the site followed by its long submergence has resulted in the preservation of substantial areas of stone-built construction. On the Central Island, for example the temple of Amun-Gereb was discovered during the side-scan sonar survey that revealed the remains of walls protruding out of the overlying sediment (Goddio 2007:75–100). Upon removal of the sands, the lines of the temenos wall of the rectangular temple enclosure could be followed, where the

Fig. 6.3 Diver investigating a collapsed wall made of limestone blocks in the interior of the Temple of Amun-Gereb (*photo* C. Gerigk © F. Goddio/Hilti Foundation)

southern wall ran for 140 m and the eastern wall 30 m. These were constructed out of large blocks (L. 0.60–1.35 m; W. 0.4–0.8 m; H. 0.4–0.7 m) of limestone, which were occasionally preserved to a height of three courses. Inside the temple compound other limestone walls were also discovered, which were sometimes preserved to five or more courses in height, indicating the position of buildings (Fig. 6.3) (Goddio 2007:79–82). The remains of wooden buildings of a post and beam type of construction were also preserved in this area of the temple compound that dated from the fourth to second centuries BC (Goddio 2007:83, Figs. 3.28–3.30). Elements of temple furniture were discovered in situ, which included the naos dedicated to Amun-Gereb (Yoyotte 2008b:309) and a large basin that was used in the rituals associated with the mysteries of Osiris (Goddio 2015:34–35), which were set in an area with the remains of paved limestone slabs (Goddio 2007:88–90).

The preservation of the temple of Amun-Gereb and other areas with substantial stone buildings, however, should not lead to the assumption that Thonis-Heracleion is some kind of Egyptian Pompeii, a city being destroyed in the midst of its life and perfectly preserved. Excavation and geophysical surveys undertaken in the local region of the north-western Delta reveal densely packed settlements with an architectural tradition that made substantial use of mudbrick. While structures made from fired bricks survive at Thonis-Heracleion, for example the circular brick base of what could be a small tholos shrine (Goddio 2007:79), mudbrick buildings do not. Even a brief comparison of the plans of Thonis-Heracleion with the nearby settlement of Naukratis (Thomas 2015), reveal the extent to which these mudbrick structures are absent from Thonis-Heracleion, with mudbrick clearly not surviving to any great extent underwater. Perhaps this is the reason why a pylon has not been discovered that would have provided the monumental entranceway for the temple of Amun-Gereb?

Another possibility for the absence of a pylon may also lie in the date of the final catastrophic sedimentary liquefaction, the evidence for which points to the eighth century AD, with the latest datable object from the site being an Abbasid dinar dated AD 785 (169 AH) (Goddio 2007:75; Bresc 2008:200–201, 352 cat. no. 432 [SCA317]). At this time, the majority of the city of Thonis-Heracleion had been largely abandoned for perhaps a millennium, with all that entails for the robbing of the architectural remains of the temples and other buildings. This action was vividly described by Eunapios (AD 345–420) in the nearby city of Canopus when in the early fifth century during the reign of Theodosius 'only the floor of the temple of Serapis they did not take, simply because of the weight of the stones which were not easy to move from the place' (Eunapius *Lives of the Philosophers*: Sopater:472). Yet, the discovery of coherent areas of buildings in Thonis-Heracleion would per- haps suggest that the city was not entirely stripped bare. More likely is that a careful selection was made of materials to recycle, or to rob, or not. An insight into this is seen in the looting of the sanctuary of Khonsu-Thoth located at the entranceway of the harbor, where the remains of 13 small limestone sarcophagi and their lids have been discovered scattered down the slope of the islet on which the sanctuary was built (Fig. 6.4) (Goddio 2015:26–28). Clearly, the looters here were more interested

Fig. 6.4 Small limestone sarcophagi for mummified ibis or falcons from the sanctuary of Khonsu-Thoth, Tonis-Heracleion (*photo* C. Gerigk © F. Goddio/Hilti Foundation)

in the contents of the sarcophagi, perhaps the mummified remains of ibis and/or falcons and any precious metals or stones that they may have contained, than in the limestone as a building material. Consequently, we should envisage that the city that collapses into the Mediterranean Sea in the middle of the eighth century AD was, on the whole, more like that of an archaeological site—the Pompeii that we see today— than a functioning seaport.

The formation of the archaeological landscape, however, is slightly more complicated and there are localized areas within Thonis-Heracleion where Pompeii-like assemblages appear to be present. Again this is due to episodes of sedimentary liquefaction, albeit on a much smaller scale than the event that eventually accounted for the entire city. A good example of this occurred in the middle of the second century BC when a large section of the temple of Amun-Gereb collapsed into the waters of the Grand Canal. Excavations within these strata have revealed the well preserved remains of utilitarian ceramics mixed in and amongst the building debris, suggesting perhaps that it was working areas of the temple that were destroyed. This collapse and the major destruction of the temple that accompanied it, appears to herald a sharp decline in the levels of occupation of the city, which is seen in a significant break in the pattern of circulation of coinage, probably in the 160s (Meadows 2015:130). A similar pattern is also observable in the ceramic record, particularly the Red Gloss wares, where it would appear that Thonis-Heracleion was largely abandoned before imports of Eastern Sigillata A into Alexandria began (i.e., before 125 BC), perhaps in the period 150–125 BC (Grataloup 2015:145).

Placing Things Beyond Reuse: Rubbish and Ritual Deposition in Waterways

Although the rise in sea level and the gentle submergence of areas of Thonis-Heracleion, alongside its large-scale abandonment, may have reduced the effects of the robbing and reuse of buildings and objects, these activities would nevertheless have resulted in extensive changes to the urban landscape. Against this, however, must be set the picture from the waterways where, removed from the processes of recycling, a very different archaeological environment exists. The research into the definition of the submerged topography of Thonis-Heracleion resulted in an extensive campaign of survey and targeted stratigraphic excavation throughout the city in areas that would have been underwater in antiquity in the port basins and the natural and artificial waterways of the landscape. In this respect the nautical assemblage discovered at the site provided key indicators that survey or excavations were most likely being conducted 'underwater'. As such, anchors were deliberately and intensively searched for in the area of the Central Port and Grand Canal, with their density and distribution helping with the interpretation of the shape of these areas. The anchors had most probably been lost in antiquity through becoming stuck fast in the silts and clays of the harbor bottom, resulting in their abandonment. Alongside these finds was also recovered the everyday detritus that forms on the bottom of an active harbor, including large quantities of imported ceramics that clearly indicate the regions with which Thonis-Heracleion was in trading contact. These demonstrate the flows of products from the Greek world, notably Corinth, the Cycladic and Ionian islands, and from Attica (Fabre 2008a:219–234; Grataloup 2008:246–252; 2010:150–154). The pottery also helps to demonstrate how the trading relationships developed over time, with imports from Corinth being plentiful during the Late Period and Athens rising to prominence towards the end of the Saïte period and during the years of Persian dominance. The pottery also indicates that in addition to products from the Greek world, Phoenician sailors were also active in Thonis-Heracleion (Goddio and Fabre 2015:151–153, 225–228, 239–245, 247, 340, and 347; Fabre and Goddio 2012).

Alongside the pottery, large quantities of objects fashioned out of lead were also recovered from these watery contexts. Elements from this assemblage are significant because of what they reveal about both rubbish disposal and the practises of ritual deposition in the city. First, the sheer scale of the lead assemblage needs to be acknowledged. Although lead appears to have been used more frequently from the Late Period onwards in Egypt, the actual numbers of objects in the archaeological record are low. This is likely to have been related to the ease with which lead objects can be recycled and the applications to which it could be subsequently put (van der Wilt 2014:37–52). At Thonis-Heracleion, however, this process of recycling does not appear to have been as prevalent as at other sites—perhaps because of its status as a major port through which much of the imported lead from mines in the Mediterranean would have passed (Fabre 2008b:342 cat. no. 357 [SCA 893]). This is clearly reflected in the archaeological record where over a thousand lead artifacts have been recovered from the excavations, many of which survived simply through their disposal in the waters of the port (van der Wilt 2014). Clearly in Thonis-Heracleion a combination of the watery landscape coupled with the abundance of the lead resulted in a set of rubbish disposal practises that removed lead from the processes of recycling and reuse that were more likely the norm in Egypt.

Not all lead objects deposited into the waters of Thonis-Heracleion were rubbish and an intriguing quantity of them also appear to have had a ritual aspect to their deposition. Van der Wilt draws our attention to a particular group of six Greek weights decorated with symbols and lettering, the combination of which indicates that they originated from Athens: the legend ΔHMO is an abbreviation of δημόσιον and it is the official guarantee for the accuracy of the weight (Fig. 6.5) (van der Wilt 2015:166–168). These weights were largely found at the western end of the Grand Canal and were deposited there, according to their lettering and the weight system in use, at a point after the second half of the fourth century BC (van der Wilt 2010). The distribution of these particular weights in a watery context stands in contrast to other weights discovered at the site, such as the light plaque weights that were concentrated on the Central Island in the area of the temple of Amun-Gereb, where there may have been a market place (van der Wilt 2015:168–169). The deposition of these rare Athenian weights in the waters of the Grand Canal—which appears to

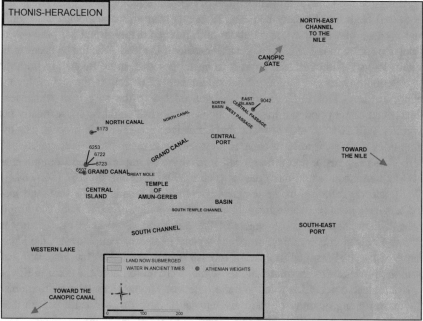

Fig. 6.5 A Greek-style half tortoise square plaque weight from Thonis-Heracleion, SCA 945 L. 5 cm, W. 4 cm, Th. 1.1 cm (*photo* E.M. van der Wilt © F. Goddio/Hilti Foundation) and a map of the site with the distribution of Athenian weights (*map* E.M. van der Wilt on F. Goddio © F. Goddio/IEASM)

have had a profound significance in the ritual topography of Thonis-Heracleion (Goddio 2015)—was surely not an act of refuse disposal but instead an act of devotion perhaps by an Athenian merchant as an offering of thanks to the Gods for their successful arrival in Egypt, or to propitiate them in advance of their return home?

The Movement of Objects and the Validity
of the Archaeological Stratigraphy

Analyses of patterns of material culture across the landscape of Thonis-Heracleion and its interpretations rests squarely upon the integrity of the archaeological stratigraphy and the extent to which the objects have remained—more or less—in proximity to their initial place of deposition. While we can be confident that the lines of walls of the temples and quaysides have not moved to any great extent, this is not to say that there has not been significant displacement of some elements of the archaeological record from their place of deposition in antiquity. Today, Aboukir Bay is extensively trawled by fishing vessels and their nets, or the remains thereof, are commonly found snagged on limestone blocks. Trawling consequently has the potential to have scrambled the archaeological remains close to the surface. For example, the four fragments of a black granodiorite sculpture of an over life-sized Ptolemaic queen (Kiss 2008:121–123, 308 cat. no. 110 [SCA 283]) were found dispersed, with farthest pieces being found 272 m from each other (Goddio 2007:28). Set against this, however, are the fragments of the monumental 6 m tall bilingual stele of Ptolemy VIII, which were found more or less in situ where they fell (Goddio 2009:ix-xvii, Figs. 2 and 5; Thiers 2008:134–137, 310 cat. no. 117 [SCA 529]; 2009). Consequently, it is clear that the extent to which trawling has redistributed objects around the site is likely to have been quite variable.

Of greater potential concern is the extent to which pieces of material culture substantially smaller and lighter than statues and architectural blocks are prone to movement within the semi-mobile marine sediments of the upper layers (cf. Muckelroy 1978:176–177). The archaeological deposits in the city and its port basins are all covered with a layer of sand of varying depths, which overlies the Nile silts and clays containing the archaeological stratigraphy with secure contexts. The interface between these layers is rich in finds and the ceramics from here are often abraded and covered in marine concretion indicating that the material has been eroded out of the archaeological layers and has undergone movement within the sands. Here it is important to note that waves of up to 4 m in height are a relatively common occurrence during winter storms, which can have a significant effect on the seabed at Thonis-Heracleion with its depth of less than 10 m. The archaeological methodology employed by the IEASM included a metal detector survey across areas of the site, which produced substantial quantities of material culture from this interface layer. The issue is, however, can the distributions of these objects around the site be trusted given the scrambling that may have occurred? Coins, which are often poorly preserved, are frequently discovered at this interface and importantly the analysis of groups of coins that were clearly once parts of hoards indicate that while there has been movement within this sandy upper layer, the distances that coins, and by implication other objects, travel may not have

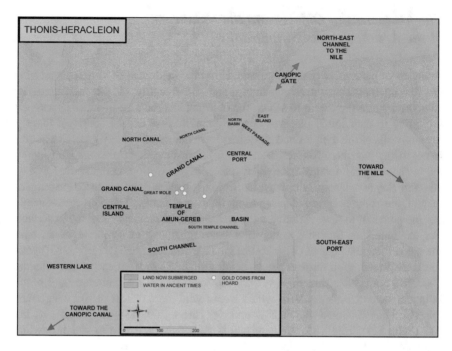

Fig. 6.6 The distribution of the five gold quarter staters of Ptolemy I from the 2004 gold hoard (*map* A. Meadows on F. Goddio © F. Goddio/IEASM)

been too large. For example, one of the hoards was composed of a group of highly diagnostic gold quarter staters of Ptolemy I that, although dispersed in the sand layer, all cluster around an original point of deposition close to the main temple (Fig. 6.6) (Meadows 2015:121–122, Fig. 6.1). This relatively limited movement of material culture, even within the mobile upper sand layers, was also demonstrated in an initial study of the pottery from around the site, which allowed the broad pattern of usage of different areas of the port city to be proposed (Grataloup 2010:156, 158). Such findings are again corroborated by an analysis of the statuettes from Thonis-Heracleion, the majority of which were discovered during the survey. Here it was found that the few statuettes that are datable by iconography also follow the same chronospatial trends as those established by the pottery (Heinz 2015:56–57). Such studies encourage us to have confidence in the distribution maps of objects presented for the site. An example of the types of analysis that can consequently be followed is a recent study of the distribution of statuettes and other cultic furniture, which has enabled the locations of small shrines and sanctuaries around the city to be proposed and investigated in more detail through stratigraphic excavation (Goddio 2015).

Summary

The first part of this chapter has presented an overview of the landscape formation processes at work in Thonis-Heracleion. It has demonstrated that there was a range of different processes in operation that together have resulted in the formation of the submerged archaeological landscape that we are investigating today. Nevertheless, this submergence was the result of a single causative factor, the human decision to locate a settlement at the edge of the 'Sea of the Greeks' on unstable Nile sediments. The initial effects of this were probably felt in the progressive inundation of the northern areas of the city, which were settled first. This was exacerbated by a series of dramatic collapses of these sediments—liquefaction—that destroyed areas of buildings, pitching them and their contents into the waterways that surrounded the city. This resulted in the creation of a number of clearly observable disaster horizons, with their tightly dated collections of material culture and building remains.

Other areas of Thonis-Heracleion, however, were abandoned at various times in its history and were substantially robbed for building materials and other goods. Some of this took place over a long period of time, with a millennium separating the abandonment of certain areas of the site and their later submergence. Alongside this, it should also be noted that the extent to which the architecture of the city is preserved is mediated by the materials out of which it was constructed. Despite their extensive robbing, the outline and foundations of the main areas of temples can be traced archaeologically due to their construction from large blocks of limestone. Whereas the houses of the great majority of the population, which were probably constructed out of far less durable materials such as mudbrick, are now lost to us.

The formation of the archaeological record at Thonis-Heracleion is also heavily influenced by the relationship of its inhabitants to the waterways and port basins into which all manner of material culture was deposited. The reasons behind individual acts of deposition are complex and probably not limited to actions such as the disposal of rubbish or placing objects in into the waters as parts of ritual acts, as well as being accidentally lost. Placing objects into water, however, effectively removed them from the practises of recycling and reuse, which has resulted in the formation of different types of archaeological assemblages compared to those originating on land. The deposition of objects into water also brings up the issue of the extent to which they would have moved from their original location around the site within the upper mobile sediments. Fortunately, analyses of the spatial location of particular groups of material culture recovered from this mobile layer have demonstrated that while there is movement of objects within it, this is relatively localized and that they are not hopelessly scrambled.

Overall then, the formation of the submerged landscape of Thonis-Heracleion can be seen to have a range of different causal factors that have effectively shaped the archaeological evidence. The next section of this chapter will turn specifically to the formation of the nautical assemblage—the ships and anchors—that have been founds in its ports and waterways and continue the investigation of this major element within the archaeological landscape.

The Formation of the Maritime Archaeological Record

The Suitability of the Hôné as a Port

When looking at the formation of the maritime archaeological record at Thonis-Heracleion, it is first necessary to consider the environmental factors and the particular set of historical circumstances that resulted in the development of a port in this location. The low-lying northern coast of Egypt was known in antiquity as being dangerous for navigation (Diodorus 1.31.2–5; Bernand 1970:22). With a general absence of offshore islands, sheltered bays, false mouths, not to mention shifting offshore sand banks, the configuration of the Egyptian maritime façade was a daunting one for the sailor (Cooper 2011, 2012a, b). The branches of the Nile were the only convenient places for anchoring and entering Egypt and of those mouths, the Canopic and the Pelusiac, were comparatively easily navigable and therefore most frequently used. Both led up the river to Memphis, the hinge between the Delta to the north and the Nile valley to the south (Chuvin and Yoyotte 1983:54; Carrez-Maratray 2005; Stanley 2007).

While the Pelusiac branch of the Nile was in use from an early date, the relative unimportance of the western part of the Delta to the Pharaonic world resulted in the lack of a 'coastal' harbor in this location until the early in the first millennium BC. Indeed, there were no purposely built ports on the shores of Pharaonic Egypt and at most there would have been just simple piers or anchorage points (Fabre 2004:19–35). The environment consisted of coastal sand dunes alongside the mouth of the river, which protected an inland area composed of marshes and lakes, interconnected by smaller rivulets. This hydrographical configuration was called a hôné by the Egyptians, which would have provided ideal conditions for a port with suitable anchorages in the lakes and the coastal dunes giving protection from the prevailing north-west winds and from north-easterly storms. The Egyptian name of the settlement, Thonis—the hôné of Saïs (Fabre 2015:180–181; von Bomhard 2012; Yoyotte 2001:25; 2008a:238 cat. no. 116)—also reveals its political links to the Saïte Dynasty of the Late Period (664–525 BC), for which the hôné was its port. The importance of the Canopic branch increases because it provided access to the developing networks of the Greeks of the central and western Mediterranean as well as the Libyans across the western desert and into Marmarica and Cyrenacia.

A coastal-riverine port was an ideal location to oversee incoming ships and guard the coast, and it was also a place of transition from two different sailing regimes. The Nile would have been full of shifting sand banks, particularly after the flood, which posed hazards that served to block passage and even cause vessels to wreck. According to figures in the *Description de l'Egypte*, year-round sailing was restricted to boats drawing less than 0.5 m of water and consequently it is likely that flat-bottomed vessels would have been necessary for the transportation of goods along the Nile and its Delta for parts of the year (Jomard 1809–1828:1, 112; Cooper 2011, 195–197). The Ahiqar scroll, a customs account written in Aramaic recording the tax dues collected at a Nile port from Ionian and Phoenician ships over the ten-month sailing season in 475 BC, gives a suggestion to how such a trans-shipment port would work. The scroll makes it clear that many of the ships from Greece arrived during the first part of the sailing season, which coincided with the period of the low Nile from March to June (Yardeni 1994; Briant and Descat 1998). If these foreign vessels drew more than half a meter of water, it is highly likely that they would have needed to trans-ship their products onto a river barge if the goods were to go up river, or else it may have required a long period at anchor in the harbors of Thonis-Heracleion awaiting easier transport conditions (Höckmann 2008–2009). In addition to these types of riverine and seagoing trading vessels, the port may also have been a naval station, as the Saïte pharaoh Nekau II is credited in Herodotus as constructing a fleet of ramming war galleys, which given the regime's propensity for employing Greek mercenaries, may have been early Greek-style banked warships (Herodotus 2.159.1; Basch 1969; Lloyd 1972; James 1992:720–723). With its location at the edge of the Mediterranean Thonis-Heracleion could have provided a base for such vessels enabling the western Delta to be protected from pirates and the incursion of hostile states, as well as the monitoring of foreign ships entering the Nile. In addition to merchant and naval vessels, we should also envisage a range of local ships and boats, including sacred vessels associated with the temples and secular ones for fishing or everyday transport of people around the waterways of the city and the local environment. A vision of the forms of such vessels that would have operated in this region is depicted on the mosaic from Palestrina, which, although dating from the Roman period in Italy, vividly illustrates a range of vessel types from papyrus boats through fishing and merchant ships to banked naval warships (Walker 2003; Fabre 2015:180).

The Ships of Thonis-Heracleion

To date, 69 ancient ships have been discovered in the waters around Thonis-Heracleion, the majority of which were found during survey work in the port basins. Upon discovery, the position of suspected vessels were marked and then later returned to for more intensive exploration, which involved the removal of the upper mobile layers of sand following the surviving edges of the hull planking where it emerged from the underlying clay of the harbor floors. It should be noted that excavation into the clay was kept to a minimum and the removal of the sand allowed the confirmation of the planking as belonging to a vessel and provided some ideas about its shape and dimensions. During this stage a plan of the ship was made and samples taken from the timbers for radiocarbon determination and wood species analysis. The initial program of research into the ships has been to simply characterize the nature of the nautical assemblage, to assess its preservation, and to develop a long-term strategy for the investigation of the different types of vessels preserved in Thonis-Heracleion. Only two vessels from the corpus have been excavated—ships 11 (Goddio and Fabre 2015) and 17 (Belov 2014a, b)—with two more—ships 43 (Robinson 2015) and 61—currently under investigation.

While the anchors found in the Central Port and the Grand Canal potentially indicate ships from the Greek and Phoenician trading worlds with examples being found in stone, wood with lead inserts in the stock, lead stocks, and also metal, they are set amongst many others that are clearly of Egyptian origin (Frost 1979; Nibbi 1991). Amongst the ships sunk in the city, however, foreign vessels appear to be largely absent (Fabre and Belov 2011:109–111, Fig. 3). Only two ships contain wood from non-native trees, pine (*Pinus* sp.), which makes up just 3% of the sampled timbers. The remainder of the vessels were constructed from woods that were either common in the Delta region, notably acacia (*Acacia* sp.), which accounts for around 78% of the sampled timbers and sycamore (*Ficus sycomorus* sp.) at 2%. Although it could have been imported, oak (*Quercus* sp.), which accounts for 7% of the wood, could also have been of Egyptian origin as according to Theophrastus (2, 4, 2, 8) and Pliny the Elder (*Natural History* 13, 63, 19) it was grown in the area of the Thebaid in the south of the country. The species of the remaining ten percent of wood samples could not be determined. The likelihood that the majority of the vessels were Egyptian and even locally produced is seen in the way in which many of the vessels were built. Egyptian boatwrights in Thonis-Heralecion used a form of carpentry for fastening together the planking that is, at present, unique to this port, which Herodotus clearly described (Herodotus Histories 2.96.; Fabre and Belov 2011:111–114; Fabre and Goddio 2013:71–73; Belov 2015).

During the initial survey to identify the vessels, they were all simply regarded as 'shipwrecks', yet with the start of more detailed excavation it can be seen that there are a much greater variety of formation processes in operation in the creation of the assemblage at Thonis-Heracleion. There are, of course, shipwrecks in the region of the site, which can be defined as vessels that have suffered catastrophic loss with them 'being broken up by the violence of the sea, or by … striking or stranding upon a rock or shoal' (Richards 2008:6–7; 2011:857). The archaeological signatures of such losses being the presence of many items that would have been in use at the time of the catastrophe, commonly cargo and objects related to the operation of the ship by its crew or for their leisure. At present it is possible to speculate on the basis of items of cargo that seem to be associated with them, that several of the vessels that have been identified in outlying areas of the port could have become stranded and wrecked in the shifting waterways, although this needs to be clarified through further investigation. In the Central Port, Ship 61, which is currently under excavation by a team from the SCA, appears to have been wrecked while it was tied up at a quayside on the Central Island close to the temple complex of Amun-Gereb. This unfortunate vessel was engulfed during an episode of sedimentary liquefaction, which saw a portion of the temple collapse, the debris from which covered and sank it.

At the western end of the Grand Canal a small 10 m long riverboat—Ship 11—was scuttled across the entranceway into the Canal by having a section carefully removed from the keel plank (Fig. 6.7). At present this particular vessel is unique in the nautical assemblage from Thonis-Heracleion and from its structure and, more importantly, from the context of its location and the artifacts deposited around it, it would appear that this boat was itself a ritual deposit (Goddio and Fabre 2015:112). The vessel was finely made out of very thin edge jointed planks of sycamore secured with mortice and tenon joincry. The choice of this type of wood is unusual and it is the only vessel from the harbor made entirely from this wood, which may have had ritual connotations as the sycamore was regarded as the 'Tree of Life' in the Mysteries of Osiris. A nautical procession associated with the festival of the Mysteries of Osiris began on the Grand Canal, in which barques similar to Ship 11 sailed between the temple of Amun-Gereb in Thonis-Heracleion and the Osirion in Canopus on the evening of the 29th of the month of Khoiak (Goddio 2015:35–39). It is consequently possible that Ship 11 was a sacred barque of the temple that was scuttled after a lifetime of work neatly at the western end of the canal directly across the channel. This was a deliberately chosen location, marking as it did the end of the Canal and the start of the Western Lake. It was also close to several structures at the edge of the waterway that have been identified as small shrines or offertories from which artifacts could be placed into the water. Consequently, the boat is

Fig. 6.7 Photo-recording Ship 11 at Thonis-Heracleion (*photo* C. Gerigk © F. Goddio/Hilti Foundation)

surrounded by many small ritual offerings that are often comprised of a stone offering plate with some food remains and a small crushed piece of lead, a metal that again has sacred associations with Osiris. Alongside these objects many *simpula*, a ladle that was used in the rituals of Osiris, were also discovered (Goddio 2015:29–34).

Wear marks on the prow of Ship 11 indicate that it was clearly a vessel that had been well used before it was deposited in the waters at the end of the Grand Canal, and its condition may well have been a reason why it was selected for disposal. At the end of the working life of a ship, its owner would have faced a choice of what to do with it: perhaps to recycle and reuse the vessel for another purpose, to salvage its wood and fittings to realize some of the economic value still retained by it, or to abandon it on some quiet part of the river (Richards 2008:61). In the vessels deposited in the Central Basin, there are most likely examples of both vessel reuse and discard, with ship recycling being difficult to identify in the archaeological record of Thonis-Heracleion. Located at the northeastern end of the Central Basin, Ship 17 provides an example of the reuse of a vessel in the creation of maritime infrastructure. The almost complete excavation of this ship revealed that it was surrounded by 14 piles (Fig. 6.8), which had been driven *c.* 2.5 m into the clays of the harbor bottom (Belov 2014a:83–86, Fig. 35). This was more than just a simple placement strategy and it is likely that the ship was used to create a jetty or wharf and that some of the piles supported some form of superstructure.

There is also evidence for what could be the abandonment and dumping of unwanted vessels at the western end of the Central Basin. The western ship graveyard, unlike the eastern graveyard discussed later in this chapter, is characterized by vessels of different type and size that appear to have been dumped here over a long period of time, being found superimposed over each other without any form of order (cf. Beattie-Edwards and Satchell 2011; Richards 2008:85, 94). This group of ships and boats would consequently be a good candidate for an abandonment graveyard.

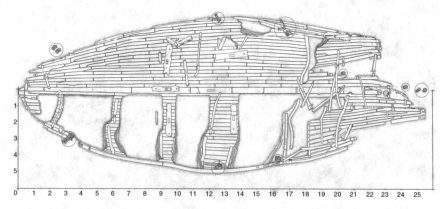

Fig. 6.8 The distribution of vertical wooden piles (*circled*) around the hull of Ship 17 from Thonis-Heracleion (*drawing* P. Sandrin © F. Goddio/IEASM)

Ship 43 and the Ship Graveyard of the Central Basin

Ship 43 was abandoned in the Central Basin alongside a number of other similar vessels in a large ship graveyard. The abandonment of this group of vessels was a deliberate and purposeful action done in the full knowledge of how it would have radically altered the local topography and the ways in which the Central Basin of the port would have been used. Consequently, through interpreting the rationale behind the creation of this abandonment graveyard we can begin to comprehend the intention and agency in the formation of the maritime landscape of Thonis-Heracleion (cf. Richards 2011:872).

Like the majority of the other ancient ships from the port city, Ship 43 was discovered during a campaign to characterize the topography of the Central Basin in 2007. This initial work allowed the rough dimensions of the vessel to be established: 23 m long by 8 m wide. A very limited number of samples were taken for radiocarbon and wood species analysis, which gave dates of 785–481 cal BC and 785–412 cal BC, with the structural elements of the vessel being made from acacia. The date for its abandonment was provided by a pile on the starboard side of the ship that was used to secure the position of the vessel (480–200 cal BC) and also from one of the pine flukes from a stone anchor of Egyptian type that was found at the prow of the ship amongst a deposit of limestone rubble ballast (405–208 cal BC). Although there are clearly too few radiocarbon dates to be precise, it would appear that Ship 43 was an old vessel at the time of its abandonment.

Ship 43 was returned to in 2011 when the current excavations were begun. This particular vessel was selected because of the opportunity that it presented to both characterize the nature of the ship graveyard through open area excavation and also because it appeared to be a similar type of vessel to Ship 17, which had previously been excavated (Belov 2014a, b, 2015). This proved to be an Egyptian riverboat—*baris*— that was most likely a multipurpose cargo boat with a wide beam and flat bottom, of a type that would have been commonly found on the river and in its Delta (Herodotus *Histories* 2.96; Fabre and Belov 2011; Belov 2014a, b, 2015). It was made with a form of construction that had hitherto never been archaeologically investigated. Consequently it seemed opportune to excavate a second *baris* and to use the data from both to undertake a full ship reconstruction for this type of vessel.

The Abandonment of Ship 43

The excavation of Ship 43 would suggest that it had undergone primary salvage prior to its abandonment, which would have followed from the departure of the crew who would have taken their personal belongings. This would have involved the removal of high value or other saleable materials, such as the sail, mast and rigging, followed by the cutting down of the deck house or other structures to reuse and recycle these materials, leaving only the hull. It is likely that this took place either on the water or with the ship grounded in the shallows and was undertaken with the explicit intention to retain the watertight integrity of the hull (cf. Buxton 1992; Buxton and Warlow 1997). Modern examples of ship breaking would indicate that little attention is paid to integrity of the hull, which is often breached at an early stage of the process: this did not occur with Ship 43 (cf. Buxton 1992:14, 28; Buerk 2006:32–34). It is consequently unlikely that recycling of the entire vessel was intended, as it is the hull that would have contained the majority of the usable wood, which was not touched. Following this stage, it was laid up in some marginal area of the port, perhaps being put into storage while other vessels were similarly prepared, with the result that it was not regularly bailed out, leading to it becoming partially submerged. The result of this is seen archaeologically in a thin layer of silt covering the bottom of the inside of the hull, which was overlain by a layer of papyrus leaves that were probably deposited as a result of the ship being swamped during the season of the inundation of the Nile.

Following this, Ship 43 was sunk alongside at least seven other vessels to create the ship graveyard in the Central Basin. It would have been floated from the shore, where it had probably received an initial ballasting of limestone (Richards 2008:156; Fabre 2012:26–27 [on ballasting]), into its required position at the western edge of the graveyard, where it was perhaps sunk and then covered with further limestone (Fig. 6.9). The rubble formed a dense mantle over the entire wreck and this deposit also continued over the stern of 43 and on to the adjacent

Fig. 6.9 Diver investigating the stem plank of Ship 43 from Thonis-Heracleion (*photo* C. Gerigk © F. Goddio/Hilti Foundation)

Ship 20, suggesting that both of these ships—and probably also the remainder of the vessels within this graveyard group—were part of the same event (Fig. 6.9).

Interpreting the Creation of the Graveyard

The abandoned ships were all old vessels, which had perhaps reached the end of their working life or were no longer worth repairing (cf. Ford 2013:212). This may suggest that the graveyard was simply the result of the disposal of unwanted vessels that had been staked down and ballasted simply to ensure that they did not move. Studies of the abandonment of unwanted vessels and the creation of graveyards indicate that the most important factor in their location is that they should never be a hazard to navigation. The port of Thonis-Heracleion was a dynamic and changing geomorphological environment where the gradual raising of sea level in the north of the city resulted in the eventual transferral of the location of the major temple to the Central Island at some point between 450 and 380 BC (Goddio 2015:45). Also the Northern Channel that led to the Nile, and which would have been the primary maritime entrance to the port, appears to have begun silting up—perhaps rapidly— in the late fifth century BC to the extent that a building was constructed in the middle of the former waterway in the fourth century BC (Goddio 2015:45). The abandonment graveyard was located between the Northern Channel and the Central Island, and as such the exact timing of the abandonment of the vessels is crucial as different interpretations rely on the northern channel being open or closed and this area of the port as being either an active zone through which ships regularly passed or a quiet backwater. If the Northern Channel was open, the ships of the eastern graveyard would have been located in a prime area for navigation and would have most likely prevented access to the anchorages in the Central Port and Grand Canal. Whereas if the Northern Channel was blocked at the time when Ship 43 was deposited, the vessels in the graveyard would not have presented a hazard. In this case, it is possible that the eastern graveyard could have been simply formed from unwanted ships being dumped here, however, this is considered unlikely as simple abandonment graveyards tend to be characterized by a haphazard collection of watercraft that accumulate over an extended period of time (Beattie-Edwards and Satchell 2011). In the eastern graveyard, the ships that have been revealed by the survey all appear to be approximately the same shape, size, be made of the same materials, and using the same construction technique (Fabre 2012; Fabre and Belov 2011). They were a carefully selected and prepared group of ships that were deposited for a clear purpose: although the question still remains what this was.

War

An appropriate context for preventing access to the Central Basin would be war, where the ships would have been scuttled carefully in position as a defensive measure designed to block a passage and deny an enemy access (Crumlin-Pederson 2002; Wood 2009). The use of ships in the creation of defensive barriers is known from the eastern Mediterranean, where from the fifth century BC onwards harbors increasingly became the focus for hostile actions, as invading forces attempted to capture coastal cities from the sea (Thucydides *History of the Peloponnesian War* 7.59.3, 69.4. Murray 2012:69–142). In Philo of Byzantium's *Compendium of Mechanics*, a treatise explaining how to attack and defend a Hellenistic city, there is a book on siege warfare (*Poliorketika*) (Marsden 1971; excerpts in Murray 2012:283–301). Philo suggests the placement of constructed platforms on the seabed covered with a mound of stones that would have created a barrier just below the waterline (Philo of Byzantium *Poliorketika* 53; Murray 2012:291–292, Fig. E.2). It is tempting to see the deposition of Ship 43 in the eastern graveyard as a part of the creation of such a permanent defensive structure with the period of the Persian invasions offering an appropriate historical context [for a consideration of the historical context within which the defences at Thonis-Heracleion may have been constructed see Fabre (2015)].

For successive waves of Persian invasions, the point of attack was almost invariably across Egypt's eastern frontier and Pelusium, with the majority of the initial military action taking place in the eastern Delta before the invading forces pushed on to secure Memphis (Ruzicka 2012). Consequently, while the Canopic mouth may have been stoutly defended, would it have been completely closed? Given the importance of Greek military aid to the Egyptians during this period, it would perhaps make strategic sense to keep open the major western harbor (Thucydides *History of the Peloponnesian* War 1.104; Diodorus Siculus Bibliotheca *Historica* 11.71.3–6). Furthermore, while the blockships would certainly have hindered enemy warships penetrating into the Central Basin and prevented them from them accessing the Central Island upon which the Temple of Amun-Gereb was located, the same island could also have been reached from the south-east port basin, which could be reached through a southern passageway to the Nile. At present there is no evidence to suggest that this entranceway was also blocked with sunken ships, although it could have been closed with more temporary lines of floating vessels linked together or chains.

The most serious complications to the defensive barrier hypothesis arise when the dating of the various elements within the landscape are taken into consideration. For while the vessels in the graveyard could have been abandoned during the uncertain times of the fourth century BC, it is at this very time that the route to the Nile that they were intended to defend was silting up. If the channel was impassable to vessels—and the construction of a building on it would certainly suggest that it was—then there would be little point in building a defensive position in this location. Consequently, we must look for other reasons behind the abandonment of Ship 43.

The Creation of Maritime Infrastructure

The second major hypothesis sees the vessels used as elements in the creation of maritime infrastructure, with them being either deliberately scuttled as part of a land reclamation scheme or used as elements in a pontoon bridge. The logical timeframe for both of these scenarios would be after the northern channel silted up. This would mean that the newly created artificial island, or structure, would not be a hazard or hindrance to ships, while a new bridge, or series of bridges, could have been used to connect the new focus of port life on the Central Island with the older parts of the city to the north.

In his *Bibliotheca Historica*, Diodorus Siculus describes a walled city at the end of the Nile "which is divided into two parts by the river and provided on each side of the mouth with pontoon bridges and guard houses at suitable points" (Diodorus Siculus Bibliotheca *Historica* 1.33.8). While only a general comment, the creation of pontoon bridges to link up the various islands and islets within the hôné would appear to be eminently sensible, particularly if they could rise and fall with the river during the period of the inundation. Herodotus' description of Xerxes' bridge across the Hellespont is a roughly contemporary account of how such a structure could have been made in Thonis-Heracleion (Herodotus *Histories* 7.36; Vegetius *De Re Militari* III.7.; Ammianus Marcellinus *Roman Antiquities* xvii 10.1, 12.4, xxvii 5.2, xxix 4.2, xxxx 5.13; Hammond and Roseman 1996). When this and other descriptions of pontoon bridges are compared to the material from the eastern graveyard there are obvious similarities and the reuse of a group of baris-type vessels as pontoons is certainly feasible. Also if some of the long piles that were used around the boats in the ship graveyard were used to allow the pontoons and the overlying roadway to rise and fall with the flood of the river, we could see this as a practical solution to the problem of providing year-round access between the central islands of the port with those in the north. The concentration of hulls would then have been created due to the collapse or swamping of this structure, or indeed its abandonment.

The other main use that wooden vessels were put to beyond their life of active service is as buildings or foundations, as part of what has been termed structural adaption, or vessel reuse (Richards 2008:861–862 [on structural adaption]; Ford 2013:198 [on vessel reuse]). It has been noted that state, or company-owned vessels were often reused in works of harbor infrastructure, as they would have presented an easily available and inexpensive resource (Ford 2013:199 [for state-owned]; Dawkes et al. 2009 [for company-owned reuse]). If Ship 43 was part of a fleet associated with the temple, it is easy to envisage a situation where the vessels were reused at the end of their working lives in the development of port infrastructure. The similarities between Ship 43 and those whose structures were adapted to help create maritime infrastructure is obvious and it would seem that the vessels of the eastern graveyard may have been reused to create a wide artificial 'island', or perhaps even the foundations for a bridge. At present it is difficult to be certain if the 'island' did indeed reach above sea level, but the graveyard could have acted as a foundation for overlying structures. Due to the limited nature of the excavations

on Ship 43 and more widely in the eastern graveyard at present there is no information on what these structures may have been (cf. Lemée 2006:119, Fig. 4.1.15).
Topographically, the land created by the vessels of the eastern graveyard could
have been used to join islands together, which were located to the east of the central
island with the temple of Amun-Gereb. Overall, given the likelihood that the
Northern Chanel would have been blocked at the time of the deposition of the ships
in the eastern graveyard, it is thought more likely that it was associated with the
creation of infrastructure, although at present exactly what this infrastructure was is
difficult to tell.

Conclusions

This chapter has investigated the entwined nature of the local environment and
human agency in the creation of the maritime landscape of Thonis-Heracleion.
Although an environmentally deterministic case could be made for the overriding
importance of nature in the creation of a suitable location for a port, or its
destruction, what we see is that there was no a priori reason for the construction of a
port here and that it was realized through social action, working in conjunction with
the environment. Furthermore, the construction of a city on the unstable
water-saturated silts of the Nile may only have hastened its end, with the localized
collapse of the temple of Amun-Gereb in the second century BC being accompanied by the apparent large-scale cessation in coin loss and ceramic use suggesting
that it was this catastrophe that triggered the abandonment of the city. Again, it is
environment and agency working in combination in the maritime landscape, albeit
this time with unforeseen consequences.

The structure of the landscape clearly helped structure the forms of rubbish
disposal and ritual deposition in Thonis-Heracleion. From the patterns of the distribution of material culture within the waters of the Nile it is obvious that the
waterways of the settlement offered a convenient place to dispose of unwanted
objects on a variety of scales from small everyday refuse to large ships. At the same
time, however, the waterways also significantly influenced and rewrote ritual
behavior. Here the use of the Grand Canal as the starting point for the ritual
navigation of the barque of Osiris during the celebration of the mysteries is clearly a
localized adaption of rituals to this environment, which are celebrated very differently elsewhere in Egypt (Goddio and Fabre 2015). These celebrations were also
accompanied by many other actions of devotion in which objects—from boats such
as Ship 11, or alternatively perhaps, elaborate models of sacred barques made of
lead, through items such as situla that were deliberately bent and thus ritually
destroyed before being deposited, to small statuettes and other votive objects—were
placed into the waters of the Nile. The recognition of such patterns of behavior is
possible thanks to the systematic and detailed programs of survey and exploration
in the waterways and the understanding that the depositional practises in these
locations are significant, structured and not hopelessly scrambled.

It is the same set of disposal practises, together with an archaeological project that has recognized the opportunity presented by the combination of ancient social action and the preservation that the environment offers, which has resulted in the location of such an exceptional number of Egyptian ships and boats from the Late through Ptolemaic periods. While our investigations are in their very early stages it is now clear that there are few shipwrecks in Thonis-Heracleion and that many more of the craft in our nautical assemblage were the result of the abandonment of vessels, whether for ritual purposes, infrastructural development, or simply for disposal. The assemblage provides clear insights into the numerous decisions about what to do with old ships enacted within a maritime landscape.

As is clear from the discussion of Ship 43, however, when looked at in detail, this chapter represents not an end point where our interpretations are secure and well founded, but a place along that journey. There are still many uncertainties with the idea that Ship 43 and the other vessels that were sunk alongside it in the eastern graveyard were all deliberately chosen and carefully prepared elements of a large-scale project of infrastructural development. The most glaring uncertainty is obviously exactly what this infrastructure was, yet by looking at the Ship 43 and the eastern graveyard in the light of the wider topographical and historical knowledge of Thonis-Heracleion, we can perhaps step back from some interpretations that no longer seem to fit with the graveyard as we currently understand it. The blockship hypothesis and the use of the graveyard as a defensive structure must consequently be set aside and instead we must seek a less dramatic and more prosaic interpretation of this major piece of port infrastructure. Here further work clearly needs to be undertaken taken to see if we can tease out precisely whether we are dealing with the remains of a land reclamation scheme to create an artificial island or perhaps the foundations for a bridge, or indeed with the remains of a bridge itself. What is clear, however, is that the research into the abandonment of the ships and boats in the nautical assemblage from Thonis-Heracleion has much to contribute to our understanding of the actions and intentions of the people who created and recreated the landscape of the port.

References

Basch, L. 1969. Phoenician oared ships. *The Mariner's Mirror* 55: 139–227.

Beattie-Edwards, M., and J. Satchell. 2011. *The hulks of Forton Lake, Gosport: The Forton Lake Archaeological Project 2006–2009*, BAR British Series 536, NAS Monograph Series No. 3. Oxford.

Belov, A. 2014a. *Études de l'architecture navale égyptienne de la Basse Epoque: Nouvelle evidence archéologique et essai de restitution en 3D*. Ph.D. thesis. Université Bordeaux Montaigne.

Belov, A. 2014b. A new type of construction evidenced by Ship 17 of Thonis-Heracleion. *International Journal of Nautical Archaeology* 43(2): 314–329.

Belov, A. 2015. Archaeological evidence for the Egyptian baris (Herodotus *Historiae* 2.96). In *Thonis-Heracleion in context*, Oxford Centre for Maritime Archaeology monograph 8, ed. D. Robinson, and F. Goddio, 195–210. Oxford.

Bernand, A. 1970. *Le Delta Égyptien d'après les texts Grecs: Les confines Libyques*. Mémoires publiés par les members de l'Institut Français d'archéologue orientale, vol. 91. Cairo.

Bowens, A. 2008. *Archaeology underwater: The NAS guide to principles and practice*. Chichester.

Bresc, C. 2008. Coins—Witness of the Islamic presence. *Egypt's sunken treasures*, ed. F. Goddio, and D. Fabre, 200–201. Munich.

Briant, P., and R. Descat. 1998. Un register douanier de la satrapie d'Égypte à l'époque achéménide (TAD C 3, 7). In *Le commerce en Égypte ancienne*, ed. N. Grimal, and B. Menu, Bibliothèque d'étude 121, 59–104. Cairo.

Buerk, R. 2006. *Breaking ships—How supertankers and cargo ships are dismantled on the beaches of Bangladesh*. New York.

Buxton, I. 1992. *Metal industries—Shipbreaking at Rosyth and Charlestown*. Kendal.

Buxton, I., and B. Warlow. 1997. *To sail no more*. Liskeard.

Carrez-Maratray, J.-Y. 2005. Réflexions sur l'accès des Grecs au littoral égyptien aux époques saïte et perse. *Topoï* 12–13: 193–205.

Chuvin, P., and J. Yoyotte. 1983. Le delta du Nil au temps des pharaons. *L'Histoire* 54: 52–62.

Cooper, J. 2011. No easy option: The Nile versus the Red Sea in ancient and medieval north-south navigation. In *Maritime technology in the ancient economy: Ship design and navigation*. ed. W. V. Harris, and K. Iara, Journal of Roman Archaeology Supplementary Series 84, 189–210. Portsmouth, RI.

Cooper, J. 2012a. "Fear god; fear the Bogaze": The Nile mouths and the navigational landscape of the medieval Nile Delta, Egypt. *Al-Masaq: Islam and the Medieval Mediterranean* 24(1): 53–73.

Cooper, J. 2012b. Nile navigation: Towing all day, punting for hours. *Egyptian Archaeology* 41, Autumn, 25–27.

Crumlin-Pederson, O. 2002. *The Skuldev ships I. Topography, archaeology, history, conservation and display*, Ships and Boats of the North 4. Roskilde.

Dawkes, G., D. Goodburn, and P. Walton Rogers. 2009. Lightening the load: Five 19th-century river lighters at Erith on the River Thames, UK. *International Journal of Nautical Archaeology* 38(1): 71–89.

Fabre, D. 2004. *Seafaring in ancient Egypt*. London.

Fabre, D. 2008a. Heracleion-Thonis: Customs station and emporium. In *Egypt's sunken treasures*, ed. F. Goddio, and D. Fabre, 219–34. Munich.

Fabre, D. 2008b. Ingot. In *Egypt's sunken treasures*, ed. F. Goddio, and D. Fabre, 342. Munich.

Fabre, D. 2012. The shipwrecks of Heracleion-Thonis: A preliminary study. In *Maritime archaeology and ancient trade in the Mediterranean*, Oxford Centre for Maritime Archaeology monograph 6, ed. D. Robinson, and A. Wilson, 13–32. Oxford.

Fabre, D. 2015. The ships of Thonis-Heracleion in context. In *Thonis-Heracleion in context*, Oxford Centre for Maritime Archaeology monograph 8, ed. D. Robinson and F. Goddio, 175–194. Oxford.

Fabre, D., and A. Belov. 2011. The shipwrecks of Heracleion-Thonis: An overview. In *Achievements and problems of modern Egyptology. Proceedings of the international conference*, September 29–October 4, Moscow. ed. G. A. Belova, 107–118. Moscow.

Fabre, D., and F. Goddio. 2010. The development and operation of the Portus Magnus in Alexandria: An overview. In *Maritime archaeology and ancient trade in the Mediterranean*, Oxford Centre for Maritime Archaeology monograph 6, ed. D. Robinson, and A. Wilson, 53–74. Oxford.

Fabre, D., and F. Goddio. 2012. Une statuette chypriote découverte à Thônis-Héracléion. In *Studies dedicated to Professor Zsolt Kiss*, Études et Travaux 25, Warsaw, 82–101.

Fabre, D., and F. Goddio. 2013. Thonis-Heracleion, emporion of Egypt, recent discoveries and research perspectives: The shipwrecks. *Journal of Ancient Egyptian Interconnections* 5(1): 1–8.

Ford, B. 2013. The reuse of vessels as harbor structures: A cross cultural comparison. *Journal of Maritime Archaeology* 8(2): 197–219.

Frost, H. 1979. Egypt and Stone anchors: Some recent discoveries. *Mariner's Mirror* 65: 137–161.

Goddio, F. 1998. Topography of the submerged Royal Quarters of Alexandria eastern harbor. In *Alexandria, The topography of the submerged Royal Quarters*, ed. F. Goddio, A. Bernand, E. Bernand, I. Darwish, Z. Kiss, and J. Yoyotte, 1–52.

Goddio, F. 2007. *Topography and excavation of Heracleion-Thonis and East Canopus (1996–2006)*. Oxford Centre for Maritime Archaeology monograph 1. Oxford.

Goddio, F. 2008. Rediscovered sites. In *Egypt's sunken treasures*, ed. F. Goddio, and D. Fabre, 26–48. Munich.

Goddio, F. 2009. Introduction. In *La stèle de Ptolémée VIII Évergète II à Héracléion*, Oxford Centre for Maritime Archaeology monograph 4, ed. C. Thiers. Oxford.

Goddio, F. 2012. Heracleion-Thonis and Alexandria, two ancient Egyptian emporia. In *Maritime archaeology and ancient trade in the Mediterranean*, Oxford Centre for Maritime Archaeology monograph 6, ed. D. Robinson, and Λ. Wilson, 121–124. Oxford.

Goddio, F. 2015. The sacred topography of Thonis-Heracleion. In *Thonis-Heracleion in context*, Oxford Centre for Maritime Archaeology monograph 8, ed. D. Robinson, and F. Goddio, 15–54. Oxford.

Goddio, F., and D. Fabre. 2015. *Osiris, Egypt's Sunken Mysteries*. Paris.

Goddio, F., D. Robinson, and D. Fabre. 2015. The life-cycle of the harbor of Thonis-Heracleion: The interaction of environment, politics and trading networks on the maritime space of Egypt's northwestern Delta. In *Harbors and maritime networks as complex adaptive systems*, ed. J. Preiser-Kapeller, and F. Daim, 25–38. Mainz.

Grataloup, C. 2008. Daily life in the Canopic region. In *Egypt's sunken treasures*, ed. F. Goddio, and D. Fabre, 246–255. Munich.

Grataloup, C. 2010. Occupation and trade at Heracleion-Thonis—The evidence from the pottery. In *Alexandria and the North-Western Delta*, Oxford Centre for Maritime Archaeology monograph 5, ed. D. Robinson, and A. Wilson, 151–159. Oxford.

Grataloup, C. 2015. Thonis-Heracleion pottery of the late period: Tradition and influences. In *Thonis-Heracleion in context*, Oxford Centre for Maritime Archaeology monograph 8, ed. D. Robinson, and F. Goddio, 137–160. Oxford.

Green, J. 2004. *Maritime archaeology: A technical handbook*. 2nd ed. London.

Hammond, N., and L. Roseman. 1996. The construction of Xerxes' bridge over the Hellespont. *Journal of Hellenic Studies* 116: 88–107.

Heinz, S. 2015. The production and circulation of metal statuettes and amulets at Thonis-Heracleion. In *Thonis-Heracleion in context*, Oxford Centre for Maritime Archaeology monograph 8, ed. D. Robinson, and F. Goddio, 55–69. Oxford.

Höckmann, O. 2008–2009. Griechischer Seeverkehr mit dem archaischen Naukratis in Ägypten. *Talanta* 40/41, 73–135.

James, T. 1992. Egypt: The twenty-fifth and twenty-sixth dynasties. In *The Assyrian and Babylonian empires and other states of the Near East, from the eight to the sixth centuries B.C. The Cambridge ancient history*, ed. J. Boardman, I. Edwards, N. Hammond, and E. Sollberger, vol. 3, 2nd ed., 677–747.

Jomard, E. 1809–1828. *Description de l'Égypte, ou recueil des observations et des recherches qui ont été faites en Égypte pendant l'expédition de l'armée française, publié par les orders de Sa Magesté l'Empereur Napoléon le Grand*. Paris.

Kiss, Z. 2008. A beauty from the depths—the dark queen. In *Egypt's sunken treasures*, ed. F. Goddio, and D. Fabre, 121–123. Munich.

Lemée, C. 2006. *Renaissance shipwrecks from Christianshavn: An archaeological and architectural study of large cargo vessels in Danish waters, 1580–1640*. Ships and Boats of the North 6. Roskilde.

Lloyd, A. 1972. Triremes and the Saïte Navy. *Journal of Egyptian Archaeology* 58: 268–279.

McKenzie, J. 2007. The architecture of Alexandria and Egypt 300 BC–AD 700. New Haven.

Marsden, E. 1971. *Greek and Roman artillery: Technical treatises*. Oxford.

Meadows, A. 2015. Coin circulation and coin production at Thonis-Heracleion and in the Delta region in the Late Period. In *Thonis-Heracleion in context*, Oxford Centre for Maritime Archaeology monograph 8, ed. D. Robinson, and F. Goddio, 121–135. Oxford.

Muckelroy, K. 1978. *Maritime archaeology*. Cambridge.

Murray, W. 2012. *The age of the Titans—The rise and fall of the great Hellenistic navies*. Oxford.

Nibbi, A. 1991. Five stone anchors from Alexandria. *International Journal of Nautical Archaeology* 20: 185–194.

Richards, N. 2008. *Ships' graveyards: Abandoned watercraft and the archaeological site formation process*. Gainesville.

Richards, N. 2011. Ship abandonment. In *The Oxford handbook of maritime archaeology*, ed. A. Catsambis, B. Ford, and D. Hamilton, 856–898. Oxford.

Robinson, D. 2015. Ship 43 and the formation of the ship graveyard in the Central Port at Thonis-Heracleion. In *Thonis-Heracleion in context*, Oxford Centre for Maritime Archaeology monograph 8, D. Robinson, and F. Goddio, 211–225. Oxford.

Robinson, D., and F. Goddio. 2015. Introduction: Thonis-Heracleion and the 'small world' of the northwestern Delta. In *Thonis-Heracleion in context*, Oxford Centre for Maritime Archaeology monograph 8, ed. D. Robinson, and F. Goddio, 1–12. Oxford.

Ruzicka, S. 2012. *Trouble in the West. Egypt and the Persian Empire 525–332 BC*. Oxford.

Stanley, J.-D. 2007. *Geoarchaeology*. Oxford Centre for Maritime Archaeology monograph 2. Oxford.

Thiers, C. 2008. Underlining the good deeds of a ruler—the stele of Ptolemy VIII. In *Egypt's sunken treasures*, ed. F. Goddio and D. Fabre, 310, 134–137. Munich.

Thiers, C. 2009. *La stèle de Ptolémée VIII Évergète II à Héracléion*, Oxford Centre for Maritime Archaeology monograph 4. Oxford.

Thomas, R. 2015. Naukratis, 'Mistress of ships', in context. In *Thonis-Heracleion in context*, Oxford Centre for Maritime Archaeology monograph 8, ed. D. Robinson, and F. Goddio, 247–265. Oxford.

van der Wilt, E. 2010. Lead weights and ingots from Heracleion-Thonis: An illustration of Egyptian trade relations with the Aegean. In *Commerce and economy in ancient Egypt: Proceedings of the third international congress for young Egyptologists 25–27 September 2009*, ed. A. Hudecz, and M. Petrik. Budapest, Oxford.

van der Wilt, E. 2014. *The place of lead in an Egyptian port-city in the Late Period*. DPhil thesis, University of Oxford.

van der Wilt, E. 2015. The weights in Thonis-Heracleion: Corpus, distribution, trade and exchange. In *Thonis-Heracleion in context*, Oxford Centre for Maritime Archaeology monograph 8, ed. D. Robinson, and F. Goddio, 161–172. Oxford.

von Bomhard, A.-S. 2012. *The decree of Saïs*, Oxford Centre for Maritime Archaeology monograph 7. Oxford.

von Bomhard, A.-S. 2015. The Stele of Thonis-Heracleion. Economic, topographic and epigraphic aspects. In *Thonis-Heracleion in context*, Oxford Centre for Maritime Archaeology monograph 8, ed. D. Robinson, and F. Goddio, 101–120. Oxford.

Walker, S. 2003. Carry-on at Canopus: The mosaic from Palestrina and Roman attitudes to Egypt. In *Ancient perspectives on Egypt*, ed. R. Matthews, and C. Roemer, 191–202, London.

Wood, L. 2009. *The bull and the barriers—The wrecks of Scapa flow*. Stroud.

Yardeni, A. 1994. Maritime trade and royal accountancy in an erased customs account from 475 BCE on the Ahiqar Scroll from Elephantine. *Bulletin of the American School of Oriental Research* 293: 67–78.

Yoyotte, J. 2001. Le second affichage du décret de l'an 2 de Nekhtnebef et la découverte de Thônis-Héracleion' Égypte. *Afrique and Orient* 24: 24–34.

Yoyotte, J. 2004. Les trouvailles épigraphiques de l'institute européen d'archéologie sous-marine dans la baie d'Abû Qîr. *Bulletin de la Société Française d'Egyptologie* 159: 29–35.

Yoyotte, J. 2008a. Stele of Thonis-Heracleion. In *Egypt's sunken treasures*, ed. F. Goddio, and D. Fabre, 236–240. Munich: Prestel.

Yoyotte, J. 2008b. Naos of the temple of Amun-Gereb. In *Egypt's sunken treasures*, ed. F. Goddio, and D. Fabre, 309. Munich.

Chapter 7
Tsunami and Salvage: The Archaeological Landscape of the Beeswax Wreck, Oregon, USA

Scott S. Williams, Mitch Marken and Curt D. Peterson

Introduction

> That a ship carrying much beeswax was wrecked here is without question. No story of the Nehalem country has ever been told without reference to it and all these are substantiated by the immense quantity of wax found scattered along the beach (Cotton 1915:46).

Unknown shipwrecks were a relatively common sight on the beaches of the Pacific Northwest coast in the nineteenth century, as recorded in accounts written in newspapers and journals of the period. Such wrecks were common enough that they typically warranted only brief mention, with the vessels usually assumed to be Japanese junks or early traders or whalers wrecked on the coast before European American settlers arrived in the mid- to late-nineteenth century. One wreck, however, drew much more attention due to its unusual cargo and a set of unique circumstances that led to the concentration, preservation, and later salvage of that cargo: the Beeswax Wreck of Nehalem, Oregon. The vessel was known as the Beeswax Wreck due to the large cargo of beeswax blocks and candles it was carrying, many of which were marked with mysterious symbols, letters, or numbers (Fig. 7.1). The Beeswax Wreck and its cargo were the topic of much scientific and secular speculation throughout the nineteenth and twentieth centuries (Stafford 1908; Cotton 1915), sparking an investigation by the United States Geological

S.S. Williams (✉)
2214 R.W. Johnson Blvd SW, Turnwater, WA 98512, USA
e-mail: willias@wsdot.wa.gov

M. Marken
ESA, 3700 East Tachevah Dr. Ste. 119, Palm Springs, CA 92262, USA
e-mail: mmarken@esassoc.com

C.D. Peterson
Geology Department, Portland State University, 1721 SW Broadway,
Portland, OR 97201, USA
e-mail: Curt.d.peterson@gmail.com

© Springer International Publishing AG 2017
A. Caporaso (ed.), *Formation Processes of Maritime Archaeological Landscapes*,
When the Land Meets the Sea, DOI 10.1007/978-3-319-48787-8_7

Fig. 7.1 A beeswax block
with a shipping symbol
carved into the surface.
Source Cotton (1915)

Survey in 1895 (*Sunday Oregonian* 1908:2) as well as two novels (Rogers 1898, 1929) and serialized fiction in regional newspapers. Learned Americans of the day had difficulty believing that a ship that predated European American settlement of the area could be large enough to carry dozens of tons of beeswax, and to what purpose.

Both the beeswax cargo and the ship's timbers, along with Chinese porcelain sherds, whole vessels, and other items of flotsam, were scattered for miles along the coast but were concentrated on the barren sandspit that separates the Nehalem River from the ocean (Fig. 7.2). Nineteenth century settlers, scientists, and curious visitors wondered what kind of ship the wreck was and where it came from, but what seemed to intrigue people the most was why it was carrying so much beeswax, and why it was spread so widely. Beeswax was found far inland, buried in fields and under the roots of trees estimated to be more than a century old (Cotton 1915). The local Nehalem and Clatsop Indians told the European American traders and settlers that the beeswax and scattered timbers were from a large ship that had wrecked "many years ago" (Coues 1897:768) before the whites settled the area.

Throughout the nineteenth century it was assumed the Beeswax Wreck ship was a supply vessel bound for the missions of Spanish California, which had been blown off course. Later twentieth century commentators suggested that the vessel was a Chinese or Japanese junk, a Portuguese merchant, or a Dutch or English pirate, based on differing analyses and research biases (Woodward 1986; Stenger 2005). Despite the Indian history, some nineteenth century observers thought the wax must be a natural deposit of mineral wax, rather than a lost cargo, simply because there was so much of it (Cotton 1915). The amount of beeswax was so great that two centuries of collection, first by local Indians and then by European American settlers, did not deplete the supply and ensured that the beeswax was known and written about extensively as the coast was settled. To this day, beeswax

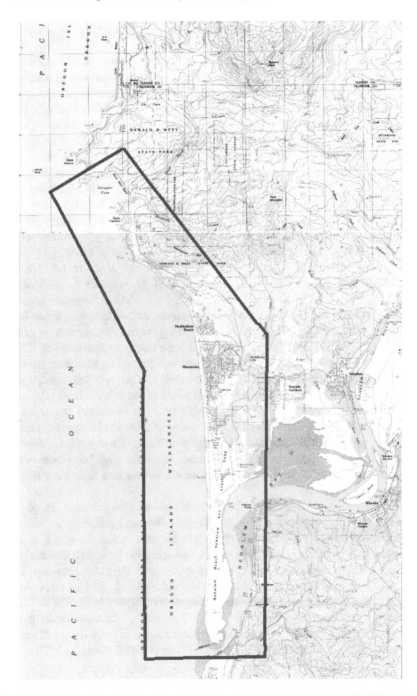

Fig. 7.2 Project location. Much of the historically reported beeswax and wreck timbers were found in the area of Nehalem Bay State Park and the town of Manzanita

and sherds of late seventeenth century Chinese export porcelain continue to be collected by local beachcombers.

As will be discussed in this chapter, the significant mechanism of deposition for the wreck debris onto the spit above the influence of storm waves and tides was a large tsunami that deposited buoyant wreck material on strandlines on the spit and the margins of the bay, creating an archaeological landscape formed first by the wrecking processes, then reshaped by the tsunami, and then reshaped again as first local Indians and then later European American settlers mined the strandlines for valued goods such as beeswax and hardwood timber. Changing environmental conditions, including tectonic uplift, sand accretion, and stabilization of the spit dunes by non-native vegetation has altered the landscape even more since the vessel wrecked and the tsunami redeposited the remains.

This chapter summarizes the historical and archaeological data supporting the identification of the Beeswax Wreck as an eastbound Manila galleon lost between 1670 and 1700, and specifically as the *Santo Cristo de Burgos* lost in 1693. The concurrence between the dating of shipwreck artifacts and the last Cascadia Subduction Zone mega-quake and tsunami in 1700 (Atwater et al. 2005) is proposed to account for the historically documented dispersal of beeswax, porcelain, and ship timbers, and the subsequent disappearance of beach wreck debris by the early twentieth century (Peterson et al. 2011). In particular, the deposition of large amounts of beeswax cargo by the tsunami onto the active aeolian dune field of the Nehalem spit is the primary reason the wreck was known to European American explorers and settlers, and is the reason the wreck was so well documented. Once deposited on the spit and bay margins, out of reach of storm waves and tides and in an area of shifting sand and migrating dunes with no obscuring vegetation, the beeswax remained accessible for discovery and collection and aroused the curiosity of passing travelers and settlers alike. If the wreck debris had not been deposited above the reach of waves and tides, the material would have been scattered and dispersed by winds and winter storms before Euro-American settlement and subsequent documentation.

The archaeological landscape of the wreck site includes not only the physical landscape and the distribution of archaeological sites and materials as they exist today, but also the historical record of eyewitness accounts of the wreck as documented in contemporary newspapers, letters, and scientific journals (Giesecke 2007; Williams 2007) combined with the physical evidence of the paleo-landscape in the form of tsunami deposits and erosion features (Peterson et al. 2011). Archaeological materials from the wreck include beeswax blocks and candles, porcelain sherds that provide a chronological framework to date the cargo, earthenware sherds, and rigging tackle and teak timbers reported to have been recovered during the mid to late nineteenth and early twentieth centuries (Rogers 1899; Giesecke 2007).

Study Area

The Nehalem Spit is located on the Pacific Northwest Coast in Tillamook County, on the north coast of Oregon (Fig. 7.2). The sand spit developed at the northern end of the Tillamook littoral cell, with long-term net transport of sand to the north that caused a large buildup of sand on the southern slope of the Neahkanie headland (Cooper 1958). The spit is 5–6 km in length and ranges from 0.5 to 1.0 km in width, and is fronted by a large foredune of 10–15 m elevation that has grown to its present height since the middle of the twentieth century due to the introduction of European dune grass to the area (Fig. 7.3). The dune grass traps migrating sand grains, causing the foredune to increase in height and to expand seaward, while at the same time stabilizing the spit surface behind the dune. A former deflation plain behind the foredune, now stabilized, decreases in height from the spit interior at circa 5 m elevation to cutbanks of circa 2 m along the bay shoreline. The narrow middle section of the spit is located adjacent to a right-angle bend in the Nehalem River, which flows behind the spit for nearly 5.0 km before exiting into the ocean.

Prior to the mid-twentieth century, the Nehalem Spit was an active dune field with little to no vegetation and dune heights of less than 8 m, but otherwise was of similar length and configuration as it is today (Fig. 7.4; Cooper 1958: Plate 2). Describing the spit in 1869 for the U.S. Coast Survey, Davidson (1869:140) wrote "between the river and the sea lies a long, narrow strip of sand dunes, having a breadth of four hundred yards and a general elevation of 25 ft." He noted that the tongue of the spit was three miles long, but at low tide the sand extended another mile to the south. In 1918 the U.S. Army Corps of Engineers completed rock jetties at the deepest part of the channel, cutting off and isolating the southern mile exposed at low tides (Fig. 7.3). This area quickly built up and became dry land, where Nedonna Beach house lots now stand.

Between the foredune and Nehalem Bay lies a low deflation basin, separated from the bay by a wooded sand ridge ending at Cronin's Point. This former basin area was graded and leveled for construction of the Nehalem Airstrip and the infrastructure and campgrounds of Nehalem State Park in the 1950s and 1960s. Prior to the vegetation planting and park facility construction, this deflation basin often became a lake in winter and low portions of it still pond water. Wreck debris and beeswax were found in this basin area as windblown dunes migrated over the spit, alternately exposing and covering wreck materials. The introduced beach grass and other non-native vegetation have since stabilized the dunes and prevented dune migration across the spit. Today, wreck materials are found in this area only rarely, usually in excavations associated with park development or maintenance in the area (Peterson et al. 2011).

Fig. 7.3 Aerial view of Nehalem Spit showing the current vegetation cover, with the jetties in the foreground and Neahkanie Mountain in the background. View to North. *Source* Beeswax Wreck Research Project, 2009

Fig. 7.4 Aerial view of Nehalem Spit taken in the 1930s looking to the Northeast, showing the lack of vegetation on the spit prior the 1950s. *Source* Beeswax Wreck Research Project Archival Photographs

Historical Accounts of the Wreck

The Lewis and Clark expedition noted that the Clatsop Indians brought beeswax to trade with the explorers when they were camped on the Columbia River in the winter of 1805–1806 (Moulton 2003). The wreck itself was first recorded in 1813 by the fur trader Alexander Henry, who noted that local Indians said the wreckage and beeswax were from a large Spanish ship wrecked many years before the fur traders settled the area in 1811 (Coues 1897:768). Henry may have referred to the ship as a Spanish vessel because of his knowledge of the trade between Manila and Acapulco, which continued until 1815, or because of the shipping marks on the beeswax blocks which included "IHS", indicating they were destined for the Catholic Church. It is possible that one of the Clatsop Indians may have shown Henry some artifact of Spanish origin from the wreck although this is not recorded in Henry's journal. Henry likely was also aware of Soto, an "old half-breed Indian" who lived up the Columbia River from Astoria and who told the fur traders in 1811 that he was the son of a shipwrecked Spanish sailor (Franchere 1854). It may have been that to Henry, the Beeswax Wreck was so obviously a Spanish ship due to the ongoing and regular trade between Acapulco and Manila and the beeswax and teak timbers scattered over the beach, that its origin was obvious and needed no further explanation. That Henry mentioned the crew was "all murdered by the natives" corresponds with stories told by Indian informants later in the nineteenth century (Smith 1899:448). Henry was surely told the same tale, indicating it is likely there were survivors of the wreck (Clarke 1899:245; Erlandson et al. 2001:49–50).

After establishment of the fur-trading post at Astoria, Oregon in 1811, there was little to no settlement of the Oregon coast until the 1840s and settlement remained very sparse from the 1840s until the 1880s. In spite of this, nearly every written account of early settlers on both the Oregon and southern Washington coasts mentions the beeswax and its association with a wrecked ship (Lee and Frost 1844; Swan 1857; Hobson 1900). Samples of the beeswax were collected in 1839 by Captain Edward Belcher during his exploration of the coast (*Overland Monthly* 1872:356; Stafford 1908:26). Writing before 1900, John Hobson tells of finding the beeswax in 1843 when he lived in the area (Hobson 1900) and the Reverend Lee and Frost (1844:107) write of the beeswax in 1844, noting:

> About thirty or forty miles to the south of the Columbia are the remains of a vessel which was sunk in the sand near shore, probably from the coast of Asia, laden, at least in part, with beeswax. Great quantities of this wax have been purchased by the Hudson's Bay Company and individuals; the writer also obtained a number of pounds of the same article from them while there, and was informed by them, that whenever the south-west storms prevail, it is driven on shore.

James Swan, living north of the Columbia River at Willapa Bay in 1852–1855, wrote of the beeswax and that the Indians said it was from a wrecked vessel. He also describes it washing ashore after "great storms" (Swan 1857:206). Davidson found beeswax on the coast in 1851, and likewise noted the Indian legend of the wreck and that "there are, occasionally, after great storms, pieces of this wax thrown ashore" (Davidson 1869:144). He goes on to say that by 1869, "formerly a great deal was found, but now it is rarely met with," although he notes that many people on the Columbia River possessed pieces of the beeswax and that he himself had seen several pieces. In the next edition of the *Coast Pilot*, Davidson (1889:453) added that the beeswax was found on the spit of the Nehalem River near its mouth after strong winds uncover it, and that the settlers "assert that part of the wreck has been pointed out by the Indians at extreme low tides."

Historic accounts report findings of more than just beeswax and timbers associated with the wreck. A large wooden tackle block was removed from the offshore wreck at an extreme low tide in 1899 (Erlandson et al. 2001), as was a small silver oil jar (Giesecke 2007). A second wood tackle block was found by a beachcomber on Nehalem beach in 1992; it is of a style typical of seventeenth century Spanish rigging blocks, and was radiocarbon dated to the seventeenth century (Erlandson et al. 2001:48). Several nineteenth century accounts mention finds of gold or silver coins from Clatsop and Nehalem beaches, or recovered from Indian burial sites, and even the recovery of Spanish gold bars (Gibbs 1971:41). Many accounts mention Chinese porcelain commonly being found on beaches in the area. Archaeological excavations in Indian house-pits and middens around Nehalem have recovered beeswax, iron, and copper artifacts as well as earthenware and porcelain sherds, some of which have been flaked into projectile points and scrapers (Woodward 1986, 1990; Scheans et al. 1990). Hobson reported finding a copper chain on the wreck (Hobson 1900). A short news article from 1881 mentions the finding of a "brass figure of the Siamese elephant" from the wreck (*Daily Astorian* 1881:3). Local fishermen reported dredging up intact blue-on-white porcelain jars and vases from deep water off Nehalem in the 1970s (Gibbs 1971:43).

The wreck and the mystery of its beeswax cargo were written about extensively in newspapers and regional journals as settlement of the region increased toward the end of the nineteenth century (Stafford 1908; Cotton 1915; Giesecke 2007; Williams 2007, 2008). Popular novels and stories were written describing the supposed adventures of the wreck survivors (Rogers 1898, 1929). When conditions were right and wreckage was exposed or beeswax was found, newspapers in Oregon and across the country carried stories about the wreck. Writers speculated on whether the beeswax was truly the cargo of a ship, and if so what its origin was, or if it was the result of a natural deposit of mineral wax (Stafford 1908; Cotton 1915).

Stafford (1908:26) notes that from the period of 1813, when the beeswax and wreck were first recorded, to 1893 "no one seems to have questioned that the deposit of wax was due to any other cause than the thing traditionally accepted as its origins—a wrecked vessel." However, as settlement of the region increased and stories of the beeswax and wreck became more common, questions were raised about how and why an ancient ship could possibly carry so much wax and how it could be so widespread: it was not only found on the beach, but was also found in the lower reaches of the river valley, buried several feet down and sometimes under the roots of trees up to 4 ft in diameter (*Sunday Oregonian* 1905:11).

A sample of Nehalem beeswax taken to the Columbian Exposition of 1893 that was cursorily examined was pronounced to be ozokerite, a mineral wax, and not beeswax. This raised the possibility to non-locals that the material was of natural rather than cultural origin. This was followed by a series of articles in the journal *Science* in 1893, where the only argument for the material being natural mineral wax (rather than beeswax from a wreck) was that it was inconceivable that a single vessel could account for such a large cargo (Stafford 1908:26). Such was the interest in the origin of the wax that the United States Geological Survey sent a geologist, Dr. J.S. Diller, to Nehalem in 1895 to determine the origin of the wax deposits. Dr. Diller concluded that the material was definitely beeswax, and not petroleum wax, and that it was from a vessel wrecked at Nehalem (Stafford 1908:29–31). Despite his findings, claims were made again in the first decade of the twentieth century that the material must be petroleum wax for no other reason than there was too much of it at Nehalem to be cargo from an ancient wrecked ship (Stafford 1908:31). Oregon newspapers of the period carried notices advertising the sale of shares in petroleum companies planning to drill for the oil they thought present in the Nehalem area (*Sunday Oregonian* 1908:8). Failure to find any oil, and the fact that the material was clearly beeswax to anyone who examined it, quietly ended the oil speculation.

In the latter half of the twentieth century archaeologists and historians became interested in the wreck (Gibbs 1971; Marshall 1984; Giesecke 2007), with Marshall (1984:178) identifying it as likely the wreck of the galleon *San Francisco Xavier*. The wreck's origin and identity were the focus of several archaeological investigations (Woodward 1986; Scheans et al. 1990). Some investigators suggested the vessel was an Asian junk, a Portuguese merchant, or a Dutch or English pirate rather than a Manila galleon (Stenger 2005; Woodward 1986). There is no historical or archaeological evidence for these claims, which are based primarily on misidentification of porcelain sherds recovered from Indian habitation sites in the area, or incomplete analysis of the archival and historical records.

Which Galleon?

The first goal of the Beeswax Wreck Project was to collect and analyze historical and archaeological information pertaining to the wreck (Williams 2007). This material included documentary evidence in the form of nineteenth and twentieth century newspaper and journal articles, specimens of beeswax blocks and ceramic sherds in museums and private collections, and reports of previous archaeological investigations in the area. From this material it is clear that the Beeswax Wreck carried goods that were typically transported by eastbound Manila galleons. The shipping symbols found on beeswax blocks over the years are Spanish shipping symbols, similar and in some cases nearly identical to shipping marks recorded in Spanish archives as those used by Spanish merchants on other Manila galleons (Mathers et al. 1988).

The sheer quantity of the wax is another clue: historic documents show that during the nineteenth century several tons of beeswax were collected from Nehalem Spit and shipped as a trade item by early settlers in the area. As noted above, the volume of beeswax was so large that several nineteenth century observers assumed it had to be natural petroleum wax, as they could not conceive of any ship carrying such a large volume of wax. These assumptions were made despite the undeniable presence of the shipping symbols carved into the beeswax blocks, the presence of candles with wicks, and even bees preserved in the wax. It is well documented that the Spanish shipped many tons of beeswax from Manila to Acapulco every year from 1565 to 1815. Spanish archives contain records of Manila galleons carrying anywhere from 60 to 100 tons or more of beeswax for use in the missions and churches of the New World during these years. No other maritime power of the era shipped beeswax in such quantities.

Dating of diagnostic Chinese porcelain from archaeological sites and beach deposits indicate a manufacturing period of 1670–1700 (Lally 2008, 2014). With artifact evidence indicating the Beeswax Wreck was a Manila galleon that wrecked sometime during or shortly after the period 1670–1700, the next step was a further search of Spanish archival documents. In the comprehensive and detailed Spanish records of galleon sailings and losses, only two Acapulco-bound galleons went missing around that time: the *Santo Cristo de Burgos*, which disappeared in 1693, and the *San Francisco Xavier*, lost in 1705 (Blair and Robertson 1909; Dahlgren 1917; Schurz 1939; Lévesque 2002). The *San Francisco Xavier* frequently has been identified as the likely candidate for the Beeswax Wreck by previous researchers (Gibbs 1971; Cook 1973; Marshall 1984; Giesecke 2007), although Erlandson et al. (2001) hypothesized an earlier galleon based on their more detailed analysis of radiocarbon dates and earlier porcelain studies. After the *San Francisco Xavier*, no other eastbound galleons disappeared, and prior to the *Santo Cristo de Burgos*, all of the missing eastbound galleons were lost in the sixteenth century: one each in 1576, 1578, and 1586. One of the sixteenth century galleons has been located in Baja, Mexico and is being investigated by the Instituto Nacional de Antropología e Historia (Von der Porten 2010).

The initial research focused on the galleon *San Francisco Xavier* as the most likely vessel for the Beeswax Wreck for two reasons. First, Schurz (1939), citing Hill (1928), stated that the *Santo Cristo de Burgos* burned and wrecked near the Marianas Islands. Second, a large tsunami is known to have struck the Oregon coast in 1700 A.D. (Atwater et al. 2005; Schlichting and Peterson 2006), and it was assumed such an event would have obliterated all evidence of a vessel that wrecked near or onshore prior to that year. It intuitively seemed more likely that the galleon of 1705 wrecked on a beach eroded by the tsunami, allowing the wreck materials to be washed onto the spit where they were historically reported.

Fieldwork began in 2007 with a terrestrial magnetometer survey of the Nehalem spit shoreline from the town of Manzanita to the river mouth five miles south. The survey was conducted during the lowest tide of the year. The purpose of the survey was to determine if any large ferrous targets such as cannons or anchors were buried in the beach, based on the hypothesis that a ship wrecked on the tsunami-eroded beach would now be inland and covered by sand redeposited on the spit since the construction of the jetties at the river mouth in the early twentieth century. Additional magnetometer survey was done in the deflation basin near the Nehalem Airstrip, in areas reported to have contained wreckage into the twentieth century (Giesecke 2007). No large ferrous anomalies were detected along the spit shore or in the deflation basin. A limited magnetometer survey was conducted with a small boat just offshore and parallel to the spit. Several potential anomalies were located, but deteriorating weather and ocean conditions prevented accurately locating the targets. Ground penetrating radar surveys were also conducted in 2007 to characterize the geomorphology of the spit and the effects of the tsunami on the landscape (Peterson et al. 2011).

Also in 2007 analysis began on a large collection of porcelain sherds collected over the previous fifteen years by a resident beachcomber (Lally 2008). The sherds were found in the surf zone, primarily in the winter, and the beachcomber recognized the sherds as potentially associated with the wreck and kept records of where each sherd was recovered. The distribution of sherds recovered from tidal and terrestrial deposits indicates an offshore source is "feeding" a beach deposit at Neahkanie Mountain, as sand moves offshore and onshore in winter and summer. Ceramic sherds also appear to be incorporated into tsunami deposits on the spit

(Peterson et al. 2011), which limits the date of their arrival in the bay to prior to the tsunami. Confirming that ceramics are incorporated into the tsunami deposit provides a *terminus ante quem* date for the wreck, as the tsunami deposit has been dated to the last large tsunami event in 1700 (Peterson et al. 2011).

Analysis of the porcelain sherds collected by the resident beachcomber was completed in 2008 (Lally 2008). This research confirmed that the cargo represented Chinese export ware intended for the markets in New Spain, as indicated by the presence of lidded coffee and hot chocolate cups and other items crafted for European rather than Asian markets. Stylistic motifs narrowed the period of manufacture of the porcelain cargo to the period between 1670 and 1700, with A.D. 1690 as the mean manufacturing date (Lally 2008, 2014).

Based on the lack of terrestrial magnetic anomalies as potential targets and the indications of an offshore source for the ceramics, the focus of the research shifted to the possibility of a pre-tsunami (i.e., pre-1700) wreck, as originally hypothesized by Erlandson et al. (2001). A pre-tsunami wreck would likely have lower hull deposits present offshore, with the historically described distribution of terrestrial wreck materials being the result of tsunami dispersion and deposition. In 2008, an additional magnetometer survey was conducted offshore from the Nehalem River mouth to Arch Cape north of Neahkanie Mountain. Results were mixed due to equipment issues, but additional targets were identified. Weather and ocean conditions prevented diving on the magnetic anomalies located that year or the ones located in 2007.

Geotechnical surveys continued through the summers of 2008 and 2009 (Peterson et al. 2011). A terrestrial magnetometer survey was conducted at Oswald West State Park north of Nehalem in 2010, to determine if anchors or other large metallic artifacts might be present there; none were found. Continued mapping of porcelain finds resulted in the identification of a likely search area for the offshore source. In late 2011, a multi-beam sonar survey identified two potential wreck targets in the area. Dive surveys to examine the two targets were undertaken in the summer of 2012, but were limited due to adverse weather conditions and no wreck materials were found. Additional magnetometer and side scan sonar surveys were undertaken on the targets in the summer of 2013, but poor visibility, rough ocean conditions, and equipment issues limited the ability to finish the systematic survey and dive on the sites. No offshore wreck deposits were located before bad weather forced an end to the dive season. In 2014, one of the targets was investigated through multiple dives, and proved to be a rock ridge rising above the sand bottom. No wreck materials were observed on the sand surface around the ridge or discovered with underwater metal detectors. The second target was not investigated in 2014, and during the 2015 dive season weather and ocean conditions again prevented examination of the target.

Geoarchaeology: The Cascadia Earthquake and Tsunami

The Cascadia Subduction Zone extends from British Columbia, Canada, through Washington and Oregon to northern California, and the Pacific Northwest is a tectonically active area that has undergone repeated mega-earthquakes and associated subsidence-generated tsunamis. The last four ruptures in northern Oregon are dated to 1700, 1300, 1100, and 300 years ago, and all are associated with near-field tsunami events (Atwater et al. 2005; Peterson et al. 2011).

The last event was an estimated magnitude 9.0+ earthquake that struck the Oregon coast in January 1700, between the years when the *Santo Cristo de Burgos* (1693) and the *San Francisco Xavier* (1705) went missing. The rupture caused widespread subsidence and generated a near-field tsunami with an estimated 8 m (over 25 ft) run-up height (Peterson et al. 2011). Given this event, understanding the physical effects of the tsunami on the shore and spit were crucial for understanding the historically reported distribution of wreck and archaeological deposits.

Spit Formation Processes

To better understand the pre- and post- tsunami evolution of the Nehalem Spit, ground penetrating radar (GPR) was used to profile the spit deposits along two axes to locate a buried subsidence erosion scarp, if present, and to determine if the Nehalem River had ever breached the spit in locations other than its historic mouth (Peterson et al. 2011). The reported presence of heavy teak timbers and beeswax blocks in the deflation plain behind the foredune led previous researchers to speculate that river originally dissected the spit, permitting wreck debris to float across the barrier of the spit foredune (Giesecke 2007). However, this would not explain the accounts of beeswax blocks and candles found up the river valley and high above the highest known storm surge lines. Several historic accounts note beeswax was found buried up the valley, or under the roots of trees, or found on a "thin, clay-like layer of sediment" within the spit (Hult 1960).

During the initial fieldwork it was noted that a large portion of the spit contained poorly sorted, water-rounded pebbles, cobbles and occasional boulders, definitely out of place in the wind-deposited dune field (Fig. 7.5). Local residents said the rocks were the remains of an abandoned rail line or road, or discarded ship's ballast, or a flood event. However, it was clear that the material was not from an abandoned road, rail line, or ship's ballast because it was rounded, poorly sorted, and too widespread. It was also clear that it does not represent a flood event, because it is too high in elevation for a flood deposit from the lower tidal reaches of the Nehalem River. The deposit was mapped and clast sizes were analyzed, revealing it as a tsunami cobble-drape deposit, composed of poorly sorted materials pulled out of the river channel and deposited on the spit surface by receding tsunami outflow (Fig. 7.6). The deposit covers an area greater than 2.25 km square and the extent

Fig. 7.5 Tsunami cobble-drape deposit exposed in a cut bank, buried under aeolian dune sand now stabilized by introduced vegetation. The tsunami layer is the layer of dark-colored sand with cobbles. *Source* Beeswax Wreck Research Project, 2009

and elevation provide data to determine the extent and elevation of the spit when the tsunami struck (Peterson et al. 2011). These data indicate the spit did not subside or erode low enough for wreckage to have been washed over the spit foredune by storm waves, and the only viable explanation to account for wreck debris on the interior spit is redeposition by a tsunami of pre-tsunami wreck materials. Rather than destroying the wreck debris, the tsunami deposited it in areas ideal for preservation and later discovery.

Current Hypothesis

Historic accounts of the Beeswax Wreck report two specific yet different wreck locations, separated by over 3 km: one wreck in the ocean at the Nehalem River mouth, and one on the deflation plain of the spit (Giesecke 2007; Williams 2007). Both locations were said to contain ship's structures, and both were only exposed during rare instances of extremely low tides and shifting sand. The initial hypothesis guiding the research project was that wreckage at the river mouth represented the location where the vessel first grounded, and the deflation plain

Fig. 7.6 Map of the tsunami cobble-drape deposit on Nehalem Spit. *Source* Peterson et al. (2011)

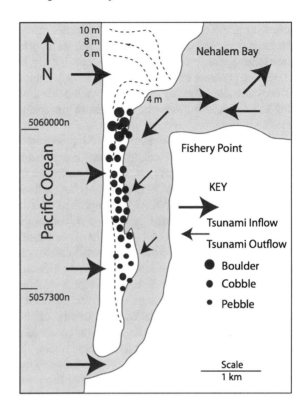

wreckage was material washed over the tsunami-eroded foredune by winter storm events. Other researchers have suggested the ship was sailing into Nehalem Bay through a former river channel at the center of the spit and became grounded, and was buried as the river mouth migrated to the south (Giesecke 2007).

There are issues with both these hypotheses. For the first, the wreck at the river mouth was described as a deck section with two mast stubs and remnant rigging tackle, while the inland wreckage was described as portions of the lower hull and scattered debris, including a large mast step, opposite the pattern that would be expected of a wreck grounding at the river mouth and then breaking up. As to the second hypothesis, there is solid geophysical evidence (Peterson et al. 2011) that the Nehalem River has entered the ocean at its historic mouth for centuries, with no evidence for any former channels to the north of the current mouth. There are also historic accounts of a third wreck off Neahkanie headland north of the spit, lying on the ocean bottom near shore and presumably in water shallow enough to be visible to someone in a passing boat. In 1894, a piece of this wreck broke free in a storm and washed ashore at nearby Arch Cape (Marshall 1984:182).

It is very unlikely that three proto-historic vessels wrecked within five miles of each other on this one stretch of North American coast. Given the tsunami history of the area and the historically reported distribution of wreckage and beeswax, and based on mapping of tsunami features across the spit, the following sequence of events is proposed to have distributed and then buried galleon wreck debris from a pre-tsunami wreck on Nehalem Spit, specifically the *Santo Cristo de Burgos* of 1693. The presence of wreck debris on the spit behind the foredune and the deposits of beeswax along the bay margins above the tide limits disqualify the later wreck of the *San Francisco Xavier* as a source of material deposited by non-catastrophic aquatic forces.

The Cascadia mega-quake of 1700 produced a large-sized tsunami run-up (~ 8 m) in the Nehalem area that swept beach and near-shore wreck debris across the spit and into Nehalem Bay. Tsunami backwash from the bay was directed down the Nehalem channel and over the low-lying sections of the spit. Shipwreck debris on the beach was remobilized and deposited over the spit by the tsunami surges, with wreck debris retained at the 6–8 m tsunami inflow strandlines and on the return flow strandlines at of 4–6 m. Coseismic subsidence during the earthquake, of about 1.5 m abrupt sea level rise, caused catastrophic beach retreat or erosion of about 150 m distance landward, over several decades after the great earthquake. Following interseismic rebound and uplift during the phase of tectonic strain accumulation, the beach began to recover, eventually burying wreck debris by the early 1900s. Given the shallow bathymetry of the Nehalem offshore area, the wreck's heavy remains including ballast, cannons, and anchors are likely located offshore and possibly survived the tsunami intact. The return of offshore sand from the 20–30 m depth range back to the beach during the last 100 years of interseismic recovery may have re-exposed offshore galleon wreck debris. This is the likely source of the ceramics washing up every year at Neahkanie Mountain. Given the orientation of the coves where the ceramics are found and the shape of the rocky headland on either side, the sherds appear to be originating from just offshore of the cove, matching the general location of the third wreck referred to in historic accounts.

If not for the tsunami and its deposition of wreck debris onto the aeolian dune field of the spit and into the bay margins above the reach of tide and storm waves, it is doubtful that enough wreck material would have survived to be documented by European American explorers and settlers. Instead, passing or fragmentary accounts of an unknown wreck lacking direct archaeological evidence would have been all that remained, similar to other areas of the Northwest Coast.

The Shipwreck Landscape

The Beeswax Wreck shipwreck is not a typical wreck site in the sense that shipwrecks are usually considered sites; that is, it is not a discrete assemblage of wreck materials in a limited location, although such a deposit may remain to be found. Instead, due to the catastrophic tsunami event that redeposited much of the wreck materials onto dry land shortly after the wrecking event, combined with nearly three centuries of actively salvaging those materials, much of the site has been removed from the primary site location and that location remains unknown. The archaeological landscape of the wreck has had both 300 years of physical formation processes acting upon it, as well as 300 years of social and behavioral landscape formation processes such as salvage and reuse. After the initial wrecking event, survivors and the local Indians salvaged wreck materials and incorporated them into their daily lives, as evidenced by artifacts recovered from nearby archaeological sites. The Nehalem Indians traded these materials to their neighbors up and down the coast, a trade that was ongoing and continued when the first European American explorers and settlers arrived more than a century after the wreck. As the Nehalem valley was settled by American immigrants, they too conducted an active trade in wreck materials, primarily beeswax but also Asian hardwood timbers, 150–200 years after the wreck. At the time, none of the later immigrants knew of the role the catastrophic tsunami of 1700 played in the distribution and preservation of wreck materials, allowing for over three centuries of collection, reuse, and wonderment of this singular occurrence of so much beeswax cargo in a limited geographical area. The result is the cognitive landscape the wreck has created: the fascination and interest the wreck has generated for two centuries of European American explorers, settlers, and adventurers, and the resulting historical record of speculative, historical, and archaeological writing about the event.

Conclusion

Based on geomorphological studies of the paleo-tsunami effects on the area landform, combined with the historic accounts of artifact distribution and the porcelain analysis, it is hypothesized that the Beeswax Wreck is the remains of the galleon *Santo Cristo de Burgos*, lost in 1693, rather than the *San Francisco Xavier* of 1705. The claim by Hill (1925, 1928) that *Santo Cristo de Burgos* burned near the Marianas, later presented as fact by Schurz (1939), is fiction, as Spanish records indicate that at least six years after the disappearance of the *Santo Cristo de Burgos* there was still no evidence of its fate (Blair and Robertson 1909; Dahlgren 1917). It is clear from historic descriptions of wreck debris dispersal and the locations where debris was historically found that normal ocean processes cannot account for the distribution of ship timbers and beeswax inland. The geomorphological study on the paleo-tsunami history and formation processes of the spit conducted by Peterson et al. (2011) provides data indicating that the tsunami of 1700 did not erode Nehalem Spit low enough to allow winter storm waves to wash a post-tsunami wreck over the spit and into Nehalem Bay, nor were there relict channels that wreckage could drift through or that a ship could sail through.

The presence of wreck ceramics in the tsunami deposit capping the spit and historic descriptions of beeswax being found on a "thin stratum of earth, like the sediment of a river freshet" (Hobson 1900:223) and under the roots of centuries-old spruce trees (*Boston Evening Transcript* 1890) "miles up the Nehalem River" (Stafford 1908:30) and hundreds of yards inland are strongly indicative of tsunami deposition. In describing the distribution of beeswax and other wreck debris, Hobson, who settled in Oregon in 1843, wrote that he believed that "some time after the wreck there was a very high freshet in the river, which spread the wax, logs and timbers all over the peninsula" (Hobson 1900:223). Hobson could not have known about paleo-tsunamis in the area, which were not recognized until the late twentieth century, and so instead concluded that a large river flood best explained the distribution of wreck materials that he witnessed.

If the Beeswax Wreck is the remains of the *Santo Cristo de Burgos*, then how to explain Schurz's statement that the vessel burned near the Marianas Islands as reported by two survivors found "years later" in the Philippines (Schurz 1939:259)? Schurz does not directly cite the source of this information, which is not reported in his original study (Schurz 1915), but he does cite Hill (1928). The story of survivors is not mentioned by Blair and Robertson (1909) or Dahlgren (1917); Dahlgren wrote regarding the *Santo Cristo de Burgos* that:

> ...it not only failed to reach port, but was wrecked, without our gaining the least knowledge of the place where that occurred. There were some suspicions that it was destroyed by fire, for at one of the Mariannes [sp.] were found fragments of burned wood, which were recognized to be woods that are found in the Philippines only. Careful search was made for many years along the coasts of South America, and in other regions; but not the least news of this ship was obtained (Bl. & Rob. [Blair and Robertson 1903–1909] XLII, p. 309).

Dahlgren's "suspicions that it was destroyed by fire" due to the finding of "fragments of burned wood" in 1917 became Schurz's definitive statement in 1939, based on the story reported by Hill (1925, 1928). Hill claimed to have found the account in archives in the Philippines, and Schurz accepted Hill's clearly apocryphal account as fact. Later researchers have continued to do so (Marshall 1984; Fish 2011). However, as late as 1699, six years after the *Santo Cristo de Burgos* sailed from the Philippines, Mexican officials reported in a letter to Spain that there was still no information about the fate of the vessel or its crew (Archivo de Indias 1699). If a document exists in the archives of the Philippines that contradicts the letter of 1699 it was not discovered by Blair and Robertson (1909), or by later researchers such as Lévesque (2002).

Percy Hill was an American expatriate living in the Philippines in the early part of the twentieth century and a prolific writer of adventure and romance stories. His account of the *Santo Cristo de Burgos* burning and two men resorting to cannibalism to survive their voyage back to the Philippines in a small boat is the opening to a tale that satirizes the Catholic Church. If survivors had been found and tried by the Catholic Church in Manila, as Hill claimed, it seems impossible that officials in New Spain would not have been aware of such an event or that Blair and Robertson (1909) would not find records of the trial in their extensive research. Instead, it is more likely that the story is fiction, as the other stories in Hill's volume clearly are, and that Hill made up the story as a plot device for his satire.

In conclusion, geoarchaeological research and landscape analysis has resulted in the development of a working hypothesis on the identity of the Beeswax Wreck and potential locations where terrestrial and marine wreck deposits are likely to be found. The effects of a catastrophic tsunami and its role in the subsequent preservation and accessibility of wreck materials has been proposed to account for the nineteenth century fascination and documentation of the wreck, which resulted in preservation of an archaeological landscape that might otherwise have been lost. In the near future, as weather and funding allow, it is planned to expand the survey area and conduct additional surveys on the offshore targets identified in 2013. If the primary site or hull remains can be located and identified, it may be possible to confirm the fate of the *Santo Cristo de Burgos*.

References

Archivo de Indias, 1699. File Filipinas 26, R.4, N18 (Documentos 1, 2, 5, 6).

Atwater, Brian F., Satoko Musumi-Rokkaku, Kenji Satake, Yoshinobu Tsuji, Kazue Ueda, and David K. Yamaguchi. 2005. *The Orphan Tsunami of 1700: Japanese Clues to a Parent Earthquake in North America*. Seattle, WA: University of Washington Press.

Boston Evening Transcript. 1890. No title. *Boston Evening Transcript*, 5 November. Boston, MA.

Blair, Emma Helen, and James A. Robertson. 1909. *The Philippine Islands, 1493-1803: Explorations by Early Navigators, Descriptions of the Islands and Their Peoples, Their History and Records of the Catholic Missions, as Related in Contemporaneous Books and Manuscripts, Showing the Political, Economic, Commercial and Religious Conditions of Those Islands from Their Earliest Relations with European Nations to the Beginning of the Nineteenth Century, Translated from the Originals*. A.H. Clark Co.: Cleveland, OH.

Clarke, Samuel. 1899. Wrecked Beeswax and Buried Treasure. *Oregon Native Son* 1(5): 245–249.

Cook, Warren. 1973. *Flood Tide of Empire: Spain and the Pacific Northwest, 1543-1819*. New Haven, CT: Yale University Press.

Cooper, William S. 1958. *Coastal Sand Dunes of Oregon and Washington*. Memoir Series of the Geological Society of America No. 72, Geological Society of America, Boulder, CO.

Cotton, Samuel. 1915. *Stories of Nehalem*. Chicago, IL: M.A. Donohue and Company.

Coues, Elliott. 1897. *The Manuscript Journals of Alexander Henry and David Thompson, 1799-1814*, vol. II. New York, NY: Francis P. Harper.

Dahlgren, Erik W. 1917. *Were the Hawaiian Islands Visited by the Spaniards Before Their Discovery by Captain Cook in 1778?: A Contribution to the Geographical History of the North Pacific Ocean Especially of the Relations Between America and Asia in the Spanish Period*. New York, NY: AMS Press (reprinted in 1977).

Daily Astorian. 1881. The City, *Daily Astorian*, 22 January: 3. Astoria, OR.

Davidson, George. 1869. *Coast Pilot of California, Oregon, and Washington Territory*. Washington, DC: Government Printing Office.

Davidson, George. 1889. *Coast Pilot of California, Oregon, and Washington*, 4th ed. Washington, DC: Government Printing Office.

Erlandson, Jon, Robert Losey, and Neil Peterson. 2001. Early Maritime Contact on the Northern Oregon Coast: Some Notes on the 17th Century Nehalem Beeswax Ship. *Changing Landscapes: "Telling Our Stories," Proceedings of the Fourth Annual Coquille Cultural Preservation Conference*, ed. Jason Younker, Mark A. Tveskov, and David G. Lewis. North Bend, OR: Coquille Indian Tribe.

Fish, Shirley. 2011. *The Manila-Acapulco Galleons: The Treasure Ships of the Pacific*. Central Milton Keynes, England: AuthorHouse UK Ltd.

Franchere, Gabriel. 1854. *Narrative of a Voyage to the Northwest Coast of America in the Years 1811, 1812, 1813, and 1814, or the First American Settlement on the Pacific*. Translated and edited by J.V. Huntington. New York, NY: Redfield.

Gibbs, James. 1971. *Disaster Log of Ships*. Seattle, WA: Superior Publishing Company.

Giesecke, Eb W. 2007. *Beeswax, Teak and Castaways: Searching for Oregon's Lost Protohistoric Asian Ship*. Manzanita, OR: Nehalem Valley Historical Society.

Hill, Percy. 1925. *Romantic Episodes in Old Manila: Church and State in the Hands of a Merry Jester—Time*. Manila, PI: Sugar News Press.

Hill, Percy. 1928. *Romance and Adventure in Old Manila*. Manila, PI: Philippine Education Co.

Hobson, John. 1900. North Pacific Pre Historic Wrecks. *Oregon Native Son*, II(5):222–224. Native Son Publishing Co., Portland, OR.

Hult, Ruby. 1960. *Lost Mines and Treasures of the Pacific Northwest*. Portland, OR: Binfords and Mort.

Lally, Jessica. 2008. Analysis of the Chinese Blue and White Porcelain Associated with the Beeswax Wreck, Nehalem, Oregon. Master's thesis, Department of Anthropology, Central Washington University, Ellensburg, WA.

Lally, Jessica. 2014. Analysis of the Beeswax Shipwreck Porcelain Collection. In *Proceedings of the 2nd Asia-Pacific regional Conference on Underwater Cultural Heritage, Hans Van Tilburg, Sila Tripati, Veronica Walker Vadillo*, ed. Brian Fahy, and Jun Kimura. Honolulu, HI: Electric Pencil.

Lee, Daniel, and J. Frost. 1844. *Ten Years in Oregon*. New York, NY: J. Collord, Printer.

Lévesque, Rod. 2002. *History of Micronesia*, vol. 20. Québec, Canada: Lévesque Publications.

Marshall, Don. 1984. *Oregon Shipwrecks*. Binford and Mort: Portland, OR.

Mathers, William M., Henry S. Parker, and Kathleen Copus. 1988. *Archaeological Report: The Recovery of the Manila Galleon Nuestra Señora de la Concepción*. VT: Pacific Sea Resources Inc.

Moulton, Guy. 2003. *The Definitive Journals of Lewis & Clark, Vol. 9, John Ordway and Charles Floyd*. University of Nebraska Press, Lincoln, NE.

Overland Monthly. 1872. About the Mouth of the Columbia. *Overland Monthly*, VIII:71–78. San Francisco, CA: John H. Carmany and Co.

Peterson, Curt D., Scott S. Williams, Kenneth Cruikshank, and John Dubé. 2011. Geoarchaeology of the Nehalem Spit: Redistribution of Beeswax Galleon Wreck Debris by Cascadia Earthquake and Tsunami (~A.D. 1700), Oregon, USA. *Geoarchaeology: An International Journal* 26(2):219–244.

Rogers, Thomas. 1898. *Nehalem, A Story of the Pacific, A.D. 1700*. McMinnville, OR: H.L. Heath.

Rogers, Thomas. 1899. Beeswax Ship is Found. *The Telephone Register*, 21 September:1, McMinnville, Oregon.

Rogers, Thomas. 1929. *Beeswax and Gold: A Story of the Pacific, A.D. 1700*. Portland, OR: J.K. Gill.

Scheans, Daniel, Thomas Churchill, Allison Stenger, and Yvonne Hajda. 1990. Summary Report on the 1989 Excavations at the Cronin Point Site (35-YI-4B) Nehalem State Park, Oregon. Report to Oregon State Parks and Oregon State Historic Preservation Office, Salem, OR.

Schlichting, Robert B., and Curt D. Peterson. 2006. Paleotsuami High-Velocity Inundation in Back-Barrier Wetlands of the Central Cascadia Margin, USA. *Journal of Geology* 114(5): 577–592.

Schurz, William L. 1915. The Manila Galleon. Doctoral dissertation, College of Social Sciences, University of California-Berkeley, Berkeley, CA.

Schurz, William L. 1939. *The Manila Galleon*. New York, NY: E.P. Dutton and Co., Inc.

Smith, Silas. 1899. Tales of Early Wrecks on the Oregon Coast, and How the Bees-Wax Got There. *Oregon Native Son* 1(7): 443–446.

Stafford, O.A. 1908. The Wax of Nehalem Beach. *The Quarterly of the Oregon Historical Society* 9: 24–41.

Stenger, Allison. 2005. Physical Evidence of Shipwrecks on the Oregon coast in Prehistory. *Current Archaeological Happenings in Oregon* 30(1): 9–13.

Sunday Oregonian. 1905. Indian Myth on Beeswax. *The Sunday Oregonian*, 18 June:11. Portland, OR.

Sunday Oregonian. 1908. All About the Beeswax of Nehalem Beach, *The Sunday Oregonian*, 26 January:2. Portland, OR.

Swan, James G. 1857. *The Northwest Coast; Or, Three Years' Residence in Washington Territory, Reprinted 1998*. Seattle, WA: University of Washington Press.

Von der Porten, Edward. 2010. Treasures Unearthed: The Archaeology of the Manila Galleon San Felipe. *Mains'l Haul* 46(1 and 2):8–15.

Williams, Scott S. 2007. A Research Design to Conduct Archaeological Investigations at the Site of the "Beeswax Wreck" of Nehalem Bay, Tillamook County, Oregon. Ms. Report to Oregon State Parks and Oregon State Historic Preservation Office. Salem, OR.

Williams, Scott S. 2008. Report on 2007 Fieldwork of the Beeswax Wreck Project, Nehalem Bay, Tillamook County, Oregon. Report to Oregon State Parks and Oregon State Historic Preservation Office. Salem, OR.

Woodward, John. 1986. *Prehistoric Shipwrecks on the Oregon Coast? Archaeological Evidence*. Salem, OR: Report to Oregon State Historic Preservation Office.

Woodward, John. 1990. Paleoseismicity and the Archaeological Record: Areas of Investigation on the Northern Oregon Coast. *Oregon Geology* 52(3): 57–66.

Chapter 8
Coastal Erosion and Archaeological Site Formation Processes on Santa Rosa Island, California

Christopher S. Jazwa

Introduction

The changes that an archaeological site undergoes after it has been abandoned by its original inhabitants can have profound effects on how it is interpreted by archaeologists. Post-depositional processes including reoccupation of the site, reuse of cultural materials, historical development, and a variety of natural processes can alter the structure and stratigraphy of sites and the provenience of cultural materials (Wood and Johnson 1978; Schiffer 1987; Erlandson 1994; Reitz and Wing 1999; Dincauze 2000; Rick et al. 2006). Some effects like burrowing by animals, cracking soils, and the growth of plant roots can influence interpretations of a site. Others, like erosion, historical development, and submergence are also much more destructive, in some cases completely destroying sites. By better understanding these processes both generally and at the local scale, we can tease out the alterations that have occurred to the archaeological record at a site since its occupation and make better recommendations for monitoring and protecting endangered sites. In the introduction, Caporaso argues that maritime landscapes are maps of human behavior in the long term. This chapter expands on this statement to demonstrate how a combination of interrelated anthropogenic and natural events can both add to and subtract from the extant maritime landscape.

This chapter is focused on the effects of coastal erosion on archaeological sites on Santa Rosa Island, the second largest of California's northern Channel Islands (NCI). The NCI include four islands, San Miguel, Santa Rosa, Santa Cruz, and Anacapa, off the coast of Santa Barbara (Fig. 8.1). The islands are an extension of the Santa Monica Mountain range and have significant topographic variation, with Diablo Peak on Santa Cruz Island reaching 740 m above sea level (Junak et al. 1995). The largest islands, Santa Cruz and Santa Rosa, have the greatest relief and

C.S. Jazwa (✉)
Department of Anthropology, University of Nevada, Reno, NV 89557, USA
e-mail: cjazwa@unr.edu

© Springer International Publishing AG 2017
A. Caporaso (ed.), *Formation Processes of Maritime Archaeological Landscapes*,
When the Land Meets the Sea, DOI 10.1007/978-3-319-48787-8_8

163

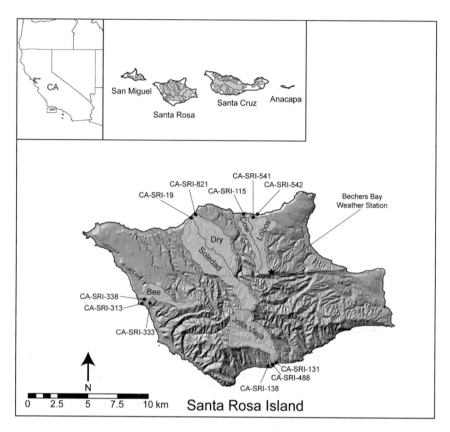

Fig. 8.1 California's Northern Channel Islands with sites investigated in this study indicated. Watersheds associated with these sites are outlined. The Bechers Bay weather station is also indicated (drawing by author, 2015)

are divided into different environmental zones with varying erosional regimes. Many of the largest sites on the NCI with the longest patterns of settlement persistence are located at or near the mouths of large drainages with reliable fresh water and access to marine subsistence resources (i.e., shellfish, fish, and sea mammals) both along the coastline and offshore (Kennett 2005; Kennett et al. 2009; Winterhalder et al. 2010; Jazwa et al. 2015b). The earliest documented settlement sites on the NCI are on Santa Rosa (Kennett et al. 2009; Winterhalder et al. 2010), and at least eight large coastal villages were distributed around the island at historic contact (Johnson 1982, 1993, 2001; Kennett 2005; Glassow et al. 2010).

This project is the start of a long-term monitoring program of coastal sites on Santa Rosa Island. In this study, I look at the effects of erosion at 11 large, permanent coastal settlement sites at four locations around the coast of the island. Focusing on environmentally diverse locations allows a better understanding of threats to site preservation. These interpretations of the rate of erosion at these sites allow archaeologists to make recommendations about locations that should be targeted for future excavation to salvage data in the future. The two primary sources of site loss include sea cliff erosion from sea level rise and marine transgression, as well as fluvial and aeolian erosion of the coastal plain and adjacent slopes. Both erosional regimes were exacerbated by historic ranching on the island during the nineteenth and twentieth centuries, at which time grazing by large herbivores denuded and destabilized the landscape. Precipitation and wind both cause gradual and punctuated erosion of archaeological landscapes, and annual erosion is influenced by weather conditions over the winter. This chapter includes a spatial analysis using GIS, along with annual site photography and monitoring of individual fixed points along erosional surfaces, to assess the condition of coastal sites and the effects of two rainy seasons (winter). This study shows that sea cliff erosion affects coastal sites around the island, with cultural materials visibly eroding out of vertical profiles in all four study locations. Another important result is that the active erosion within sites, beyond that occurring along sea cliffs, is most strongly related to the geographic location of the sites. The strongest effects are to those sites along the northwest and north coasts of the island, which are directly in the path of northwesterly winds and storms. Winter weather patterns also influence the amount of erosion. Heavier rain during the 2014–2015 winter than during 2013–2014 is reflected in more significant erosion at the sites monitored in this study.

NCI Archaeological Record

California's NCI have a rich archaeological record, in part because of the relatively few disturbances to sites on the islands compared to the adjacent mainland (Erlandson 1984; Johnson 1989; Rick et al. 2006). Unlike the mainland coast, which has been subject to over 150 years of historical development, the islands have been protected for much of this time as ranches, and recently as part of the National Park Service and the Nature Conservancy. Furthermore, the NCI have well-preserved stratigraphy at many archaeological sites because the islands were never colonized by burrowing pocket gophers, which minimizes damage to site stratigraphy from bioturbation (Erlandson 1994; Rick et al. 2006). There is strong evidence that people were visiting and seasonally exploiting the resources available on the NCI by at least 13,000 years ago (Johnson et al. 2002; Kennett 2005; Erlandson et al. 2007, 2008, 2011; Kennett et al. 2008). This long record of occupation has been subjected to a variety of post-depositional processes, the effects of which accumulate over time.

The rich archaeological record from the NCI has been used to address a broad variety of anthropological questions. The islands are unique because of the range of research topics that archaeologists address, despite their relatively small size (Jazwa and Perry 2013:1). Broad theoretical issues that have been addressed on the NCI range from population of the Americas and early colonization (Erlandson 1994, 2002; Erlandson et al. 2007, 2011; Braje et al. 2013) to the development of complexity among hunter-gatherers (Johnson 1982, 1993; Arnold 1992, 2001a, b; Kennett 2005; Kennett et al. 2009; Perry and Delaney-Rivera 2011). Recent work has included the antiquity of marine adaptations (Erlandson et al. 2011; Braje et al. 2013), paleoethnobotany (Gill 2013, 2014; Gill and Erlandson 2014), and ritual on the islands (Perry 2013), among many other topics. One of the common archaeological themes addressed on the NCI has been settlement strategies and how they developed over time from the earliest explorers through the complex population of hunter-gatherer-fishers that lived on the islands at the time of European contact in the sixteenth century (Kennett 1998, 2005; Perry 2003; Kennett et al. 2009; Winterhalder et al. 2010; Gusick 2012, 2013; Glassow 2013; Jazwa et al. 2013).

Although recent research has highlighted the importance of plant resources in their diet (Timbrook 1993, 2007; Gill 2013, 2014; Gill and Erlandson 2014), the inhabitants of the NCI had a strong maritime focus, subsisting heavily on marine resources. For this reason, many of the permanent settlement sites on the islands were coastally oriented (Kennett 2005; Kennett et al. 2009; Winterhalder et al. 2010; Jazwa et al. 2013, 2015b; Jazwa 2015), putting them at risk of loss from marine transgression. Prior to about 10,000 years ago, sea levels were lower than at present ($\sim$70–85 m at 13,000 years ago), and the four islands were all connected to form one landmass, Santarosae (Kennett et al. 2008; Watts et al. 2011; Reeder-Myers et al. 2015). As a result of eustatic sea level rise, the islands were separated by 9000 years ago (Kennett et al. 2008; Reeder-Myers et al. 2015). This rise in sea level caused a 65% decrease in land area and may have destroyed and/or submerged evidence of early permanent settlement (Kennett et al. 2008). Therefore, we have an incomplete record of the earliest inhabitants of the NCI. The earliest date for human activity on the NCI is Arlington Man, a partial skeleton dated to between 13,000 and 12,900 cal BP (Johnson et al. 2002; Agenbroad et al. 2005). More than 50 shell midden sites have been found on the NCI dating to the terminal Pleistocene and early Holocene (before $\sim$7550 cal BP; Erlandson et al. 2011), and recent surveys of marine terraces have demonstrated that there are more to be found (Rick et al. 2013). However, many of these sites are in locations that are protected from rising shorelines, either inland or at high elevation (Rick et al. 2001, 2013; Erlandson et al. 2008, 2011).

The first confirmed evidence for permanent residential bases on the NCI dates to the middle Holocene ($\sim$7550–3600 cal BP; Kennett 2005; Winterhalder et al. 2010; Jazwa et al. 2015a, b), shortly before sea levels stabilized around 6000–5000 cal BP (Cole and Liu 1994; Rick et al. 2005; Rick 2009). If earlier sites of this kind existed, they may now be underwater. The highest concentration of large middle Holocene residential bases was along northwest coast of Santa Rosa Island (Jazwa et al. 2015a, b). These sites are heavily eroding partially because in the Santa Barbara Channel region, prevailing northwesterly winds carry weather systems that soak the west and northwest coasts of the NCI. These coasts are much wetter, foggier, and cooler than the more protected southern sides of each island (Hochberg 1980; Junak et al. 1995; Fischer et al. 2009).

During the late Holocene, (after $\sim$3600 cal BP), there was an increase in the number and geographic range of residential bases on Santa Rosa Island (Kennett et al. 2009; Winterhalder et al. 2010; Jazwa et al. 2015b; Jazwa 2015). This was related to wetter conditions that increased the habitability of previously unoccupied locations (Jazwa et al. 2016). The number of permanent settlement sites continued to increase until the Medieval Climatic Anomaly (1150–600 cal BP), a period of extreme drought conditions throughout the United States southwest (Stine 1994; Raab and Larson 1997; Jones et al. 1999; Yatsko 2000; Kennett 2005; Jones and Schwitalla 2008). After this time, settlement contracted to a smaller number of large village sites (Jazwa 2015; Jazwa et al. 2015b). Despite their young age, large Late Period (650–168 cal BP) village sites have been damaged by erosion and other factors discussed in the next section.

Processes Influencing NCI Sites

Rick et al. (2006:571, Table I) summarize the taphonomic and formation processes affecting archaeological sites on the NCI. Natural threats include movement of materials by animals, argilliturbation, aeolian processes, faunalturbation, floralturbation, fluvial processes, marine processes, and mass wasting/gravity. Cultural processes include prehistoric human activities, historical impacts, and the introduction of exotic animals. Many of the natural processes accumulate over time, either at relatively constant rates or during punctuated events like storms. Muhs (1987:566) suggested that most California sea cliffs retreat at average rates between about 0.01 and 0.05 m/year, although much of the retreat occurs rapidly during storms. The relative effects of these processes should differ based on a number of factors, including site location (e.g., coastal vs. interior, windward vs. leeward), topography (e.g., slope), and nearby features (e.g., canyons, ridgelines, peaks, streams). Climatic changes, like droughts (Raab and Larson 1997; Kennett 2005; Jazwa et al. 2016), changing El Niño frequencies (Kennett 2005), and changes in fire regimes (Kennett et al. 2007, 2008), along with natural or pre-ranching, human-induced changes in the fauna in the habitat (Arnold 2001b; Agenbroad et al. 2005; Rick et al. 2006, 2008), could influence these rates.

Cultural processes not only affect the archaeological record directly, but they can significantly exacerbate the rate at which natural processes alter archaeological sites. Despite the fact that the NCI were never developed to the extent of the southern California mainland, the historic occupation of the islands had an extreme impact on archaeological sites, primarily through the introduction of grazing animals. At least 19 animals, most notably large herbivores, were introduced to the islands (Schoenherr et al. 1999; Rick et al. 2006:580, Table III). Among the most destructive of these were sheep, (*Ovis aries*), pig (*Sus scrofa*), deer (*Odocoileus hemionus*), and cattle (*Bos Taurus*). All have since been removed from the NCI because of efforts by Channel Islands National Park and the Nature Conservancy, but decades of overgrazing have had major deleterious effects on archaeological sites and island morphology (Johnson 1980; Arnold 2001b; Rick et al. 2006).

The stripping of vegetation from the landscape made the soil more susceptible to coastal, fluvial, aeolian, and other sources of erosion on the islands. This chapter is focused on erosion of coastal sites, both along sea cliffs and further inland (Fig. 8.2).

Fig. 8.2 The two main types of coastal erosion. **a** Sea cliff erosion resulting from marine transgression at CA-SRI-131 along the south coast; **b** Erosion along the edge of an eroding stream cut at CA-SRI-2 along the northwest coast (photos by author, 2013)

The topography of the coast of Santa Rosa Island consists of two flat landforms in most locations, particularly near the mouths of the major drainages. All of the sites in this study are along the coastal plain, which is raised between 5 and 25 m above sea level, and coastal beaches and are typically at the same elevation of the canyon bottoms. This has protected the sites from marine transgression to some degree, but has left them vulnerable to erosion along sea cliffs (Fig. 8.2a). While people were living in these locations, the raised landform would have been protected from flooding. The next landform is a coastal terrace at the elevation of the ridgelines, typically several hundred meters above the coastal plain. There is a steep slope (>20°) between these levels, which has experienced significant slumping and erosion. Coastal shell midden sites are common along both of these landforms, but because of erosion, it is difficult to determine how many sites have been lost in this way, or what percentage of these sites have been lost.

Many large coastal sites contain erosional gullies running through them, significantly increasing the length of erosional banks. This is particularly evident in sites like CA-SRI-2 (Fig. 8.2b), and two sites included in this study, CA-SRI-19 and CA-SRI-115. While it is clear that each of these sites is being damaged along sea cliffs, the biggest threat to these sites is from gullies. At some sites, most notably CA-SRI-19, the top strata of much of the site area is eroded away, revealing cultural material among denuded surfaces.

The four locations chosen for this analysis are in different regions of the island and are each affiliated with a drainage or pair of drainages. These regions and the 11 sites included in this study vary in age, condition, and location (Table 8.1). CA-SRI-19 and -821 are at the mouth of Dry and Soledad Canyons along the northwest coast of the island. Both sites are in very poor condition, with most of their area affected by erosion (Fig. 8.3). The coast is oriented to the northwest in this part of the island, so it is the part of the island most affected by wind and storms. CA-SRI-115, -541, and -542 are at the mouth of Cow and Lobos Canyons, along the north coast of the island. They are also in the path of northwesterly storms and wind. CA-SRI-131, -138, and -488 are at the mouth of Jolla Vieja Canyon along the south coast of the island, and are therefore protected from northwesterly weather systems by the interior mountain ridge of the island. Although there are erosional gullies along the south coast of the island, the primary threat to the three sites in this area is sea cliff erosion (Fig. 8.4). Finally, CA-SRI-313, -333, and -338 are at the mouth of Bee Canyon along the west coast of Santa Rosa Island. This location is also somewhat protected from northwesterly weather systems.

Methods

Weather Monitoring

Hourly instrument data from a weather station at Becher's Bay in northeastern Santa Rosa Island (Fig. 8.1) have been collected by Channel Islands National Park

Table 8.1 Sites included in this study with descriptive information

Site	Location	Monitored points	Site area (m²)	Confidence	Condition	To the coast Distance (m)	Direction	Erosion source	Primary occupation period
CA-SRI-19	Dry/Soledad	3	56,425	Medium	Very poor	0	NW	Sea Cliff, Drainage, Arroyo	Middle-late holocene
CA-SRI-821	Dry/Soledad	1	2814	Medium	Very poor	0	NW	Sea Cliff, Drainage, Arroyo	Middle holocene
CA-SRI-115	Cow/Lobos	2	18,687	Medium	Poor	0	N	Sea Cliff, Drainage, Arroyo	Middle-late holocene
CA-SRI-541	Cow/Lobos	1	271	High	Good	~160	N	Arroyo	Late holocene (late period)
CA-SRI-542	Cow/Lobos	1	461	Low	Poor	0	N	Sea Cliff	Late holocene
CA-SRI-131	Jolla Vieja	1	7360	Medium	Medium	0	S	Sea Cliff, Arroyo	Late holocene
CA-SRI-138	Jolla Vieja	2	8036	Low	Medium	0	S	Sea Cliff, Arroyo, Road Cut	Middle-late holocene
CA-SRI-488	Jolla Vieja	1	1715	Low	Poor	0	SE	Sea Cliff	Late holocene
CA-SRI-313	Bee	2	6734	Low	Very Poor	0	S	Sea Cliff, Arroyo	Late holocene
CA-SRI-333	Bee	1	853	High	Good	~150	SW	Arroyo	Late holocene (late period)
CA-SRI-338	Bee	1	3578	Low	Poor	~500	W	Drainage	Middle holocene

Fig. 8.3 Overview photos of the sites in this study along the northwest (CA-SRI-19 and -821) and north (CA-SRI-115, -541, and -542) coasts of Santa Rosa Island (photos by author, 2013)

since April 1990.[1] I incorporated hourly and daily data from this database for precipitation and wind from August 1, 2013 through July 31, 2015. I compiled the total monthly and annual (August–July) precipitation during this range, along with the number of days and individual hours with heavy rain (>1 mm) or storms (>5 mm). Similarly, I calculated the daily average wind speed and maximum wind gust for each month and each year (August–July), along with the number of days each month with average wind speed over 10 m/s and maximum wind gusts over 20 m/s.

---

[1]These data are available online from the Western Regional Climate Center of the Desert Research Institute (http://www.wrcc.dri.edu).

Fig. 8.4 Overview photos of the sites in this study along the south (CA-SRI-131, -138, and -488) and west (CA-SRI-313, -333, and -338) coasts of Santa Rosa Island (photos by author, 2013)

Site Monitoring

For this project, my team and I used a Trimble Geo7X handheld GPS unit to generate maps of each of the 11 sites. Using the Trimble system, we created polygons of the overall site distributions and extent of surface erosion along with polylines of each of the erosional banks. We collected these data over the course of three summers (2013–2015), but the precision of the Trimble system was not adequate to demonstrate significant annual differences in the erosion at these sites. Future long-term data collection should reveal changes at this scale. Data on the extent of surface erosion were supplemented using aerial photos taken in 2014 that are available in ArcGIS 10. These distributions provide a way to compare the threats from erosion between the sites. I used the proportion of the surface area that is eroding in each site (eroding surface area divided by overall site surface area) and the ratio between the length of erosional banks and the overall surface area (erosional bank length divided by overall site surface area) of the site as indicators of active erosion and the potential for continued loss of cultural material at each of the sites.

Point Monitoring

My team and I chose 16 specific points on the landscape along erosional banks to monitor annual erosion, with four from each geographic area. For Dry and Soledad Canyons, there are three monitoring points at CA-SRI-19, a very large site, and one at CA-SRI-821. For Cow and Lobos Canyons, there are two points at CA-SRI-115 and one each at CA-SRI-541 and CA-SRI-542. For La Jolla Vieja Canyon, we placed two points at CA-SRI-138 and one each at CA-SRI-131 and CA-SRI-488. And for Bee Canyon, we placed two points at CA-SRI-313 and one each at CA-SRI-333 and CA-SRI-338. At each chosen location, we placed a metal stake in the ground set 1 m back from the erosional bank. These points were chosen to be adjacent to 25 cm × 25 cm units that I excavated previously (Jazwa 2015) in part because radiocarbon dates for the deposition of cultural materials were available for these locations. Monitoring points were placed approximately 20 cm to the side of each excavation unit. Stakes were initially placed during July of 2013, and their distance from the erosional bank was again measured during May–June 2014 and August 2015 to monitor annual change.

Results

Annual Weather

There were clear differences in weather patterns between 2013–2014 and 2014–2015, both related to rainfall and wind. Overall, 2014–2015 was wetter than 2013–2014 in both total rainfall and frequency of heavy rains. During 2013–2014, there was a total of 173.3 mm of precipitation, with 18 days and 37 h with at least 1 mm of rainfall, and 8 days and 6 h with at least 5 mm of rainfall (Table 8.2). The rainy season began in November, which was the first month to have more than 10 mm of total precipitation and more than 5 mm during a single day (14.2 mm) and hour (5.6 mm). However, this was largely limited to a single day and rainfall was minimal during December and January. The peak in precipitation during 2013–2014 occurred in February, with 93.0 mm of rainfall and 6 days with more than 5 mm. Heavy rainfall continued through March and April, and there were no days with more than 1 mm of rainfall between May and July.

Table 8.2 Precipitation data for the Becher's Bay weather station on Santa Rosa Island

Year	Month	Total precipitation (mm)	Daily			Hourly		
			At least 1 mm	At least 5 mm	Maximum (mm)	At least 1 mm	At least 5 mm	Maximum (mm)
2013	August	6.1	1	0	1.0	0	0	0.3
	September	2.3	0	0	0.5	0	0	0.3
	October	6.6	1	0	3.3	1	0	2.3
	November	17.8	2	1	14.2	4	1	5.6
	December	4.3	2	0	2.0	1	0	1.1
2014	January	1.3	0	0	0.5	0	0	0.3
	February	93.0	6	6	28.5	24	4	12.9
	March	19.6	3	1	10.7	3	1	6.6
	April	10.2	3	0	4.8	4	0	2.0
	May	3.0	0	0	0.5	0	0	0.3
	June	6.1	0	0	0.5	0	0	0.3
	July	3.3	0	0	0.5	0	0	0.3
2013–2014	August–July	173.3	18	8	28.5	37	6	12.9
2014	August	2.8	0	0	0.5	0	0	0.3
	September	3.5	0	0	0.8	0	0	0.3
	October	17.0	1	1	15.2	2	1	12.5
	November	7.4	2	1	5.1	2	0	3.1
	December	188.2	8	4	89.2	39	11	12.7
2015	January	15.5	2	1	11.7	4	0	1.5
	February	17.8	3	2	7.1	3	0	1.8
	March	2.3	0	0	0.5	0	0	0.5
	April	4.8	2	0	2.5	2	0	2.0
	May	8.1	1	1	5.3	2	0	2.8
	June	11.2	1	7	5.8	1	0	2.6
	July	4.8	0	3	0.8	0	0	0.3
2014–2015	August–July	283.2	20	11	89.2	55	12	12.7

Data are compiled monthly and annually (August–July) for 2013–2014 and 2014–2015

The period 2014–2015 had higher precipitation than the previous year with 283.2 mm. Additionally, the rainy season began and peaked earlier than in 2013–2014. There was 17.0 mm of precipitation during October 2014, but it was mostly (15.2 mm) confined to a single day, and largely 1 h (12.5 mm). Precipitation peaked during December with 188.2 mm, which was more than the entire previous year. There were 8 days and 39 individual hours with more than 1 mm, and 4 days and 11 individual hours with more than 5 mm. This suggests that there were intense storms during December 2014 that had the potential to accelerate erosion of archaeological sites on the islands. Rains continued through January and February, with minimal rain from March through July except for individual heavy rain days in May and June.

Unlike precipitation, wind was stronger during 2013–2014 than 2014–2015 (Table 8.3). The daily average wind speed during 2013–2014 was 6.5 m/s, with a maximum daily average of 14.9 m/s. The highest average daily wind speeds were from February through May, with the highest monthly average in May of 8.6 m/s. The highest wind gusts were also from February through May. Although the highest average daily maximum wind gust of 16.8 m/s occurred in May, the annual maximum of 30.9 m/s occurred in February, coincident with the peak in rainfall. There were 53 days with average wind speed over 10 m/s and 35 days with gusts over 20 m/s. Both were most highly concentrated in May, with 13 and 9 days,

Table 8.3 Wind data for the Becher's Bay weather station on Santa Rosa Island

Year	Month	Daily average wind speed			Daily maximum wind gust		
		Average (m/s)	At least 10 m/s	Maximum (m/s)	Average (m/s)	At least 20 m/s	Maximum (m/s)
2013	August	6.8	3	10.6	13.0	0	18.8
	September	6.6	3	13.0	14.0	3	22.8
	October	5.4	1	10.4	12.1	0	19.7
	November	5.2	2	11.6	12.2	1	20.1
	December	5.6	4	12.3	12.6	4	21.9
2014	January	5.5	2	10.6	12.9	1	24.6
	February	7.1	6	13.4	15.7	5	30.9
	March	6.9	4	12.9	15.2	4	25.9
	April	7.1	6	13.2	14.9	4	22.4
	May	8.6	13	14.9	16.8	9	25.5
	June	7.1	6	11.7	14.3	3	21.9
	July	5.7	3	11.9	11.9	1	21.0
2013–2014	August–July	6.5	53	14.9	13.8	35	30.9
	August	5.4	0	9.3	11.5	1	21.5
	September	6.9	2	12.4	13.4	2	22.8
	October	5.0	1	10.2	11.1	1	20.1
	November	5.4	1	12.6	12.3	1	22.4
	December	6.4	3	12.2	14.9	8	28.6
2015	January	4.1	0	7.9	10.1	0	16.5
	February	5.7	3	12.7	12.7	2	22.8
	March	5.5	3	13.2	12.8	3	22.8
	April	7.1	6	12.4	15.1	5	24.6
	May	7.9	6	12.5	15.6	5	21.5
	June	6.7	7	11.2	13.8	1	21.5
	July	6.0	3	12.6	12.3	0	19.7
2014–2015	August–July	6.0	35	13.2	13.0	29	28.6

Data are compiled monthly and annually (August–July) for 2013–2014 and 2014–2015 respectively. Overall, 2014–2015 was less windy than the previous year, and all of the corresponding measurements were lower. The daily average wind speed for the year was 6.0 m/s, with a maximum of 13.2 m/s. The windy season was February through May, with the maximum monthly average wind speed of 7.9 m/s during May. The annual maximum wind gust of 28.6 m/s occurred in December, which was coincident with the heavy rainfall. The highest average daily maximum wind gust of 15.6 m/s occurred in May, although the value of 14.9 m/s for December was uncharacteristically high for that month. There were 35 days with average wind speed over 10 m/s and 29 days with gusts over 20 m/s. The former was most highly concentrated in May, with 6 days, and the latter was most highly concentrated in December, with 8 days. Heavy wind and precipitation during December 2014 maximized the potential for erosion.

Site Condition Assessment

The spatial data collected by the Trimble system and aerial photographs was used to interpret the relative threat of erosion to each of the sites in this study (Table 8.4). The first measure that I use is an "erosion index," which I am defining as the ratio between the length of erosional banks and the overall site area. In this study, this ranges from 1.6 at CA-SRI-19 to 16.4 at CA-SRI-333. This provides an approximation of how much of the site is exposed in vertical profile by erosion and therefore at risk of complete loss. The second measure is the proportion of the site by surface area that is actively eroding, including the surface area that is denuded with cultural materials that have clearly been moved by post-depositional processes. In this study, this ranges from 0.04 at CA-SRI-541 to 0.56 at CA-SRI-19. This provides a measure of how much of the site has been influenced by erosion.

Table 8.4 Calculated erosion index and proportion of the surface area of the site that has been eroded

Site	Location	Site area (m^2)	Erosion length (m)	Erosion index	Erosion area (m^2)	Proportion of site
CA-SRI-19	Dry/Soledad	56,425	912	1.6	31,744	0.56
CA-SRI-821	Dry/Soledad	2814	164	5.8	1194	0.42
CA-SRI-115	Cow/Lobos	18,687	742	4.0	5597	0.30
CA-SRI-541	Cow/Lobos	271	15	5.7	12	0.04
CA-SRI-542	Cow/Lobos	461	67	14.4	224	0.49
CA-SRI-131	Jolla Vieja	7360	171	2.3	601	0.08
CA-SRI-138	Jolla Vieja	8036	461	5.7	1301	0.16
CA-SRI-488	Jolla Vieja	1715	96	5.6	202	0.12
CA-SRI-313	Bee	6734	149	2.2	3277	0.49
CA-SRI-333	Bee	853	140	16.4	206	0.24
CA-SRI-338	Bee	3578	241	6.7	1552	0.43

The sites at the mouth of Dry and Soledad Canyon, CA-SRI-19 and -821, have been heavily influenced by erosion, both along the coastline and because of water flow, wind, and gravity (Fig. 8.3). However, CA-SRI-19 has the lowest erosion index of all of the sites in the study (1.6). This number is small because it only takes into account erosional banks along the edges of gullies. This is a large site with wide erosional gullies. Although there is a greater length of erosional banks at CA-SRI-19 than any of the other sites in the study, they are over a much larger area. The extensive erosion at this site is much more apparent when considering the proportion of the site with surface erosion (0.56). More than half of the site by area is actively eroding, and even areas that are today grass covered and stabilized may have eroded in the past. CA-SRI-821, which is on an actively eroding block of sediment to the east of Dry Canyon, has a higher erosion index (5.8). Midden is visible both around the edge of the site eroding into deep gullies and in a series of erosional banks throughout the site. The site is on a steep slope, and much of it is actively eroding, which is reflected in the fourth highest proportion of the site that is actively eroding among all sites (0.42).

The sites at the mouth of Cow and Lobos Canyon are also heavily eroded. Like CA-SRI-19, CA-SRI-115 has a low erosion index because it is large and has wide gullies (4.0). The proportion of the site area that has been eroded is lower than at CA-SRI-19 (0.30). Most of the damage to the site has been done along sea cliffs and a series of deep gullies cutting through the site. CA-SRI-542, a smaller site, has also been heavily eroded, primarily along the sea cliff. For this reason, it is difficult to determine how much of the site has been lost, and even how far it extends into the coastal plain. The site has a relatively high erosion index (14.4) and a high proportion of what remains of the site has been eroded (0.49), because all of the cultural materials are exposed along the sea cliff, which runs along the length of the site. CA-SRI-541, which is associated with a small rockshelter, is further inland and has been less damaged, with no sea cliff erosion at the site. The erosion index (5.7) represents a single bank in which dense midden deposits are exposed. Most of the site is intact, with a low proportion that has been eroded (0.04).

The sites at the mouth of Jolla Vieja Canyon contain large intact areas, with most of the erosion along sea cliffs. CA-SRI-488 is a small site that is eroding heavily along the sea cliff (index 5.6). Like CA-SRI-542, it is difficult to determine the original extent of the site. CA-SRI-131 and -138 are morphologically similar to each other, and both are on a coastal plain that is largely intact within the sites except for heavy erosion along the sea cliff. The erosion index at CA-SRI-138 (5.7) is larger than CA-SRI-131 (2.3), mostly because of a road cut running through the northeast corner of the site. In both cases, erosion comprises a relatively small proportion of the surface area of the site (0.16 for CA-SRI-138; 0.08 for CA-SRI-131). At CA-SRI-131, a crack running parallel to and approximately 1 m from the sea cliff suggests a large block of sediment will soon break off into the ocean. This crack has widened over the course of the three years of this study.

There is large variability in the condition of sites at the mouth of Bee Canyon. CA-SRI-333 is a relatively small site not along the sea cliff. A shallow gully cuts through the middle of the site, exposing midden in a small rise. The gully runs the length of the site, which contributes to its high erosion index (16.4). All of the surface area of the site that is eroded is within this gully, and an intermediate proportion of the site has been eroded (0.24). CA-SRI-313 is eroding heavily, with large parts of the densest midden in imminent danger of being lost. The erosion index (2.2) is low because of the wide gullies at the site. However, a large proportion of the surface area of the site has been eroded (0.49). Finally, CA-SRI-338 is heavily eroding into the deep canyon cut in the south wall of Bee Canyon. Like CA-SRI-542, the erosion index is high (6.7) because the top of the canyon wall with eroding material runs along the length of the site. It is difficult to determine how far the site extends into the coastal plain. Midden is eroding extensively down the slope, suggesting that a large proportion of the surface area of the site has been eroded (0.43).

Point Measurements

Data from individual points within each of these sites provide more nuanced perspectives of annual erosion of cultural materials along the coast (Fig. 8.5; Table 8.5). During the first year, 2013–2014, only Point 1 of CA-SRI-115 experienced more than 3 cm of erosion. 16 cm of the wall of the gully was lost at that

Fig. 8.5 Monitored points during May–June 2014 after one year, with the adjacent excavation units highlighted. Note the significant erosion at CA-SRI-19 and CA-SRI-115, and minimal erosion at CA-SRI-541 and CA-SRI-333 (photos by author, 2014)

Table 8.5 Annual measurements of the distance from the established points to the erosional banks and qualitative interpretations of the annual degree of erosion at the 16 monitored points at the sites in question

Site	Point	Location	Facing	Location	Year 1 (2013) Distance	Year 2 (2014) Distance	Erosion degree	Year 3 (2015) Distance	Erosion degree
CA-SRI-19	1	Dry/Soledad	N	Sink Hole/Sea Cliff	100	100	Low	95	Medium
CA-SRI-19	2	Dry/Soledad	E	Sink Hole	100	99	High	94	High
CA-SRI-19	3	Dry/Soledad	N	Sink Hole	100	98	Medium	97	Medium
CA-SRI-821	1	Dry/Soledad	N	Erosional Bank	100	99	Low	96	Medium
CA-SRI-115	1	Cow/Lobos	W	Deep Gully	100	84	Very High	80	Very High
CA-SRI-115	2	Cow/Lobos	N	Sea Cliff	100	98	Medium	97	Medium
CA-SRI-541	1	Cow/Lobos	W	Shallow Gully	100	100	Low	100	Low
CA-SRI-542	1	Cow/Lobos	N	Sea Cliff	100	99	Medium	97	High
CA-SRI-131	1	Jolla Vieja	S	Sea Cliff	100	99	Low	98	Low
CA-SRI-138	1	Jolla Vieja	NE	Road Cut	100	100	Low	99	Low
CA-SRI-138	2	Jolla Vieja	S	Sea Cliff	100	98	Medium	96	Medium
CA-SRI-488	1	Jolla Vieja	SE	Sea Cliff	100	99	Low	98	Low
CA-SRI-313	1	Bee	S	Sea Cliff	100	99	Medium	92	Medium
CA-SRI-313	2	Bee	W	Sea Cliff	100	100	Low	100	Low
CA-SRI-333	1	Bee	N	Shallow Gully	100	100	Low	100	Low
CA-SRI-338	1	Bee	NW	Sink Hole/Canyon	100	100	Low	97	Medium

location. In all, the most extensive erosion during that year was along the northwest and north coasts of the island. The only unit of the eight sites from those two locations with no noticeable evidence of erosion during that time span was CA-SRI-541, which is further inland and more protected from coastal erosion than any of the others. There were clear indications of erosion at CA-SRI-19, particularly at Point 2, where fresh erosion was clear in sinkholes near the unit. There was also evidence for erosion at Point 2 of CA-SRI-115 along the sea cliff. Erosion was less pronounced along the south and west coasts of the island. The excavation units from these locations were largely intact, with the exception of CA-SRI-138, Point 2 along the south coast and CA-SRI-313, Point 1 on the west coast.

During the second year (2014–2015), erosion accelerated at all four of the major locations around Santa Rosa Island included in this analysis. It was particularly pronounced at the sites along the north and northwest coasts. Point 1 at CA-SRI-115 continued to erode, outpacing all other units with 20 cm lost in total. All four units from the mouth of Dry and Soledad Canyons have experienced at least 3 cm of erosion, as did all of the units from the mouth of Cow and Lobos Canyons besides those at CA-SRI-541. This site is further inland and protected from coastal storms, so erosion there has been minimal. In addition, there is evidence for undercutting at several of these points. There has been more extensive erosion of sediment than reported in Table 8.5, but sediment at the top of the unit is held in place by vegetation. Undercutting is most extreme at CA-SRI-542, with the lower levels eroded 17 cm further than the upper stratum. At SRI-115, Point 2 along the coast, there is 5 cm of undercutting. Even CA-SRI-541, which is protected and has experienced minimal erosion overall, has 3 cm of undercutting. At the mouth of Dry and Soledad Canyons, there is 3 cm of undercutting at CA-SRI-821 and 3 cm of undercutting at Point 1 at CA-SRI-19. Points 2 and 3 at CA-SRI-19 have experienced significant erosion, but there is no undercutting because there is no vegetation in the upper strata.

At Jolla Vieja Canyon, the monitored points continued to erode at the same rate as during 2013–2014, with Point 2 at CA-SRI-138 experiencing the greatest erosion at 2 cm/year (4 cm total). All four points experienced at least some erosion during 2013–2015. At Bee Canyon, on the other hand, neither CA-SRI-333 nor Point 2 at CA-SRI-313 experienced measurable erosion. Erosion at Point 1 of CA-SRI-313 (8 cm) and at CA-SRI-338 (3 cm) was more substantial. There is no clear evidence for undercutting at the sites at either Jolla Vieja or Bee Canyons.

Discussion

This study highlights the importance of understanding post-depositional process for reconstructing human activity at archaeological sites and landscapes in coastal settings (Rick et al. 2006). Overall, erosion seems to be most closely related to the geographic location of sites on the island, including their distance from the coastline. The sites along the northwest and north coasts of the island are more eroded than those along the south and west coasts. All sites directly along the coast have some evidence of sea cliff erosion. However, fluvial and aeolian erosion appear to be most pronounced along the northwest coast of the island, which is directly in the path of northwesterly storms.

The age of sites appears to be a much less important factor for their preservation than their geographic location (Table 8.1). Although middle Holocene sites like CA-SRI-19 and -821 along the northwest coast have experienced substantial erosion, contemporaneous sites like the early locus at CA-SRI-138 on the south coast are better preserved and only exposed by a recent road cut. Conversely, the west coast site, CA-SRI-313, which postdates 1300 cal BP, has been eroded significantly and the cultural materials there are at risk of being completely lost. Additionally, the more recent late Holocene deposits at CA-SRI-131 and the later locus at CA-SRI-138 are eroding much more quickly than the earlier locus at CA-SRI-138. These patterns are primarily the result of the destabilization of the sediment associated with historic grazing and devegetation (Johnson 1980; Rick et al. 2006; Arnold 2001b). Erosion subsequent to these events overtook any differences in gradual erosion between older and younger sites, with the exception of the effects of marine transgression.

Rising sea levels have submerged large areas of the NCI. Although sites postdating the slowing of sea level rise at 5000–6000 cal BP are prominent along the coast, it is clear that they are still being affected by rising sea levels. The gradual submergence of coastal sites is a destructive process that could limit whether sites could be found underwater (Fig. 8.2a). Therefore, when searching for submerged sites, it is important to consider not only the locations that would have been optimal for occupation, but also for preservation during submergence (Stright 1990; Jazwa and Mather 2014). Although the destabilization of the sediments in archaeological sites by devegetation accelerated the erosion along sea cliffs, their eventual loss or submergence would be inevitable. Furthermore, unlike weather, which has a greater impact on the northwest and north coasts of the NCI, sea level rise affects the entire island.

The formation of erosional gullies, arroyos, and sinkholes was exacerbated by the devegetated landscape. Sites along the northwest coast of the island have been damaged most extensively by this form of erosion. Both CA-SRI-19 and -821 have been heavily affected, with extensive gullies and erosional surfaces throughout the sites. CA-SRI-19 has extensive erosional surfaces, and the large percentage of the surface area of the site that has been eroded reflects the imminent risk to the site. The uppermost strata for large parts of the site have been eroded away, making it difficult to find locations that were intact when selecting excavation units. Both CA-SRI-19 and -821 are on steep slopes and contain parts that are heavily slumping. Other similar sites further west along the coast, like CA-SRI-2 (Fig. 8.2b), are also heavily eroded.

The coastal plain along the north coast near Lobos and Cow Canyons has also been heavily degraded. CA-SRI-115 is a large site with a similar erosional pattern to CA-SRI-19 overall, although more of the site is intact. CA-SRI-541 is largely intact, but it is part of a group of sites on the coastal plain directly to the west of the mouth of Lobos Canyon. CA-SRI-116, the largest site there, is morphologically similar to CA-SRI-115 and there is heavy erosion along the coastal plain between these two large sites. CA-SRI-542, on the eastern side of the mouth of Lobos Canyon, is almost completely lost along the sea cliff and not enough is left to assess the other effects of erosion. Active erosion at these sites is evident in the annual spot checks for erosion near the excavation units, with the erosional bank near Unit 1 of CA-SRI-115 eroding more quickly than any of the others.

There is a different pattern near the mouth of Bee Canyon. The coastal plain there is largely intact, with a series of both deep and shallow gullies dissecting the landscape. With the exception of sites like CA-SRI-313 that has midden visible in the sea cliff, sites in this area are only visible where they are exposed in these gullies. I speculate that CA-SRI-333 is among a series of buried sites in the coastal plain that have yet to be found. Additionally, an old ranch road suggests that these gullies are the result of recent erosion. Tire tracks show that it was formerly possible to drive along the coastal plain, although today the gullies are cut too deeply for them to be crossed safely (Fig. 8.6). At CA-SRI-313, midden is visible in small pockets throughout the site and the densest deposits including Point 1 are in extreme danger of being lost during a single event. CA-SRI-338 is also eroding heavily into Bee Canyon itself, although the intact deposits are very dense and it is unclear how much of the site remains.

Near the mouth of Jolla Vieja Canyon, there is pronounced erosion along the shoreline. Erosional gullies surround all three sites and their effects on the sites in

Fig. 8.6 A recently formed erosional gully along the coastal plain of western Santa Rosa Island, south of Bee Canyon. Note the old road that runs through the gully but is no longer passable (photo by author, 2015)

this analysis are minimal. Like the west coast, there is little erosion at the surface of the sites, so they are more stratigraphically intact than those along the northwest coast of the island. There is a road cut in the northeast corner of CA-SRI-138 where Point 1 was placed. This locus appears stable, although the top 40 cm of sediment is much softer than the bottom 45 cm. Therefore, there has been much more erosion and slumping of the upper strata of this location since the road was constructed.

Slumping and erosion of the upper coastal terrace and slope caused some movement and redeposition of cultural materials on sites on the lower coastal plain. At SRI-19, cultural material can be found eroding down this slope, which is less severe than in other locations around the island. Jazwa et al. (2015b) noted two radiocarbon reversals from materials collected from sites in this study. An early radiocarbon date from the upper cultural stratum of Unit 1 (late Holocene) at this site likely reflects cultural material from elsewhere in the site eroding downslope. There is a similar radiocarbon reversal in the uppermost stratum of Unit 1 from CA-SRI-138 (middle Holocene). This suggests that caution should be used in interpreting chronologies from eroding coastal sites on the NCI.

More nuanced interpretations about the role of weather in coastal erosion are available when comparing the observed annual weather patterns with the measured erosion at the 16 excavation units. The winter of 2014–2015 was more conducive to site erosion than the winter of 2013–2014. Not only was there more than 50% more rain overall during the second year, but there were also more large storms. There was more rainfall during December 2014 alone than during the entire previous winter, including the days with the highest rainfall (December 2; 89.2 mm rainfall, 20.1 m/s wind gust) and wind gusts of the year (December 11; 4.83 mm rainfall; 28.61 m/s wind gust). This was reflected in accelerated annual erosion at Lobos/Cow, Dry/Soledad, and Bee Canyons, as measured at individual monitored points.

Future work should include continued data collection to better understand the specific effects of post-depositional process on individual archaeological sites on Santa Rosa Island. This will benefit greatly from a long-term data set for the 11 sites in this study. I plan to revisit each of the sites once per summer with the Trimble system to assess changes in the extent and location of the erosional banks and surface exposures of eroded material. I will also continue to monitor the 16 specific points and photograph the sites. By using the Trimble and site photographs to look at large scale changes and augment this with individual points for small-scale changes, it is possible to get a clearer picture of long-term changes at coastal sites on Santa Rosa Island. By continuing to collect data and compare these annual changes with fluctuations in weather over a longer time, it will be possible to make predictions about how sites will respond in different conditions in the future.

Conclusion

This study looks at the effects of post-depositional processes at coastal sites and landscapes, particularly those related to marine processes. I focus here on role of marine transgression and weather on the erosion of 11 coastal sites in different locations around Santa Rosa Island. The damage to cultural resources caused by both factors was exacerbated by historical ranching that included grazing by sheep, cows, pigs, deer, and elk. This left the landscape denuded, unstable, and vulnerable to other post-depositional processes. The removal of all of these species from the NCI has led to a recovery in vegetation across the islands (Rick et al. 2006; Corry and McEachern 2009; Glassow et al. 2010), potentially restabilizing the landscape and mitigating some of the coastal erosion at archaeological sites. One of the central goals of the long-term monitoring started here is to assess the effects of these changes on future taphonomy of these coastal sites.

The most important factor influencing coastal erosion between locations is where they are located, and not age of the sites, largely because of the extensive erosion during the twentieth century. Since site formation is primarily dependent on location rather than age, coastal erosion affects the formation of the landscape spatially. Potential for understanding how the physical landscape was negotiated by past inhabitants of Santa Rosa Island has therefore been altered. Understanding these physical taphonomic processes may allow archaeologists to reconstruct theoretical site use for the portions of the landscape that have been lost by erosion or submergence. Exceptions to the observation that erosion is largely age-independent are those sites that were occupied before about 5000–6000 years ago when there was a decrease in the rate of sea level rise. It is likely that many sites occupied prior to that time have been submerged. By understanding the effects of coastal processes and marine transgression at coastal sites, it is possible to better reconstruct the processes that sites that are now underwater underwent before being submerged. This could help predict the location and potential condition of sites on the continental shelf. This study also helps understand which coastal sites are at risk of being damaged and lost, and can assist with future decisions about research and management of archaeological resources on the NCI and elsewhere.

Acknowledgments I would like to thank Channel Islands National Park, including Kelly Minas, Don Morris, and Ann Huston, for help with this project. This research was supported by Channel Islands National Park (135414, P11AC30805), the National Science Foundation (BCS-1338350), and Pennsylvania State University. Sarah Mellinger, Kyle Garcia, Michelle Wilcox, Amber Marie Madrid, Terry Joslin, Blaize Uva, Hugh Radde, Stephen Hennek, Nathan Beckett, Henry Chodsky, Mike Price, and Kyle Jazwa assisted with fieldwork. I would also like to thank Alicia Caporaso for inviting me to be a part of this volume and for her helpful comments on this manuscript. Ben Ford also provided helpful comments on a draft of this paper.

References

Agenbroad, Larry D., John R. Johnson, Don Morris, and Thomas W. Stafford Jr. 2005. Mammoths and Humans as Late Pleistocene Contemporaries on Santa Rosa Island. In *Proceedings of the Sixth California Islands Symposium*, ed. Dave K. Garcelon, and Catherin A. Schwemm, 3–7. Arcata, CA: Institute for Wildlife Studies.
Arnold, Jeanne E. 1992. Complex Hunter-Gatherer-Fishers of Prehistoric California: Chiefs, Specialists, and Maritime Adaptations of the Channel Islands. *American Antiquity* 57(1): 60–84.

Arnold, Jeanne E. 2001a. The Chumash in World and Regional Perspectives. In *The Origins of a Pacific Coast Chiefdom: The Chumash of the Channel Islands*, ed. Jeanne E. Arnold, 1–20. Salt Lake City: University of Utah Press.

Arnold, Jeanne E. 2001b. Social Evolution and the Political Economy in the Northern Channel Islands. In *The Origins of a Pacific Coast Chiefdom: The Chumash of the Channel Islands*, ed. Jeanne E. Arnold, 287–296. Salt Lake City: University of Utah Press.

Braje, Todd J., Jon M. Erlandson, and Torben C. Rick. 2013. Points in Space and Time: The Distribution of Paleocoastal Points and Crescents on the Northern Channel Islands. In *California's Channel Islands: The Archaeology of Human-Environment Interactions*, ed. Christopher S. Jazwa, and Jennifer E. Perry, 26–39. Salt Lake City: University of Utah Press.

Cole, Kenneth L., and Geng-Wu Liu. 1994. Holocene Paleoecology of an Estuary on Santa Rosa Island, California. *Quaternary Research* 41: 326–335.

Corry, Patricia M., and A. Kathryn McEachern. 2009. Patterns in Post-Grazing Vegetation Changes among Species and Environments, San Miguel and Santa Barbara Islands. In *Proceedings of the Seventh California Islands Symposium*, ed. Christine C. Damiani, and David K. Garcelon, 201–214. Arcata, CA: Institute for Wildlife Studies.

Dincauze, Dena F. 2000. *Environmental Archaeology: Principles and Practice*. Cambridge: Cambridge University Press.

Erlandson, Jon M. 1984. A Case Study in Faunalturbation: Delineating the Effects of the Burrowing Pocket Gopher on the Distribution of Archaeological Materials. *American Antiquity* 49: 785–790.

Erlandson, Jon M. 1994. *Early Hunter-Gatherers of the California Coast*. New York: Plenum Press.

Erlandson, Jon M. 2002. Anatomically Modern Humans, Maritime Adaptations, and the Peopling of the New World. In *The First Americans: The Pleistocene Colonization of the New World*, ed. Nina Jablonski, 59–92. San Francisco: Memoirs of the California Academy of Sciences.

Erlandson, Jon M., Torben C. Rick, Terry L. Jones, and Judith F. Porcasi. 2007. One if by Land, Two if by Sea: Who Were the First Californians? In *California Prehistory: Colonization, Culture and Complexity*, ed. Terry L. Jones, and Kathryn A. Klar, 53–62. Landam, MD: Altamira Press.

Erlandson, Jon M., Torben C. Rick, Todd J. Braje, Alexis Steinberg, and René L. Vellanoweth. 2008. Human Impacts on Ancient Shellfish: A 10,000 Year Record from San Miguel, California. *Journal of Archaeological Science* 35: 2144–2152.

Erlandson, Jon M., Torben C. Rick, Todd J. Braje, Molly Casperson, Brendan Culleton, Brian Fulfrost, Tracy Garcia, Daniel A. Guthrie, Nicholas Jew, Douglas J. Kennett, Madonna L. Moss, Leslie Reeder, Craig Skinner, Jack Watts, and Lauren Willis. 2011. Paleoindian Seafaring, Maritime Technologies, and Coastal Foraging on California's Channel Islands. *Science* 441: 1181–1185.

Fischer, Douglas T., Christopher J. Still, and A. Park Williams. 2009. Significance of Summer Fog and Overcast for Drought Stress and Ecological Functioning of Coastal California Endemic Plant Species. *Journal of Biogeography* 36: 783–799.

Gill, Kristina M. 2013. Paleoethnobotanical Investigations on the Channel Islands: Current Directions and Theoretical Considerations. In *California's Channel Islands: The Archaeology of Human-Environment Interactions*, ed. Christopher S. Jazwa, and Jennifer E. Perry, 113–136. Salt Lake City: University of Utah Press.

Gill, Kristina M. 2014. Seasons of Change: Using Seasonal Morphological Changes in *Brodiaea* Corms to Determine Season of Harvest from Archaeobotanical Remains. *American Antiquity* 79(4): 638–654.

Gill, Kristina M., and Jon M. Erlandson. 2014. The Island Chumash and Exchange in the Santa Barbara Channel Region. *American Antiquity* 79(3): 570–572.

Glassow, Michael A. 2013. Settlement Systems on Santa Cruz Island Between 6300 and 5300 BP. In *California's Channel Islands: The Archaeology of Human-Environment Interactions*, ed. Christopher S. Jazwa, and Jennifer E. Perry, 60–74. Salt Lake City: University of Utah Press.

Glassow, Michael A., Todd J. Braje, Julia G. Costello, Jon M. Erlandson, John R. Johnson, Don P. Morris, Jennifer E. Perry, and Torben C. Rick. 2010. *Channel Islands National Park Archaeological Overview and Assessment*. Ventura, CA: Cultural Resources Division, Channel Islands National Park.

Gusick, Amy E. 2012. Behavioral Adaptations and Mobility of Early Holocene Hunter-Gatherers, Santa Cruz Island, California. Doctoral dissertation, Department of Anthropology, University of California, Santa Barbara.

Gusick, Amy E. 2013. The Early Holocene Occupation of Santa Cruz Island. In *California's Channel Islands: The Archaeology of Human-Environment Interactions*, ed. Christopher S. Jazwa, and Jennifer E. Perry, 40–59. Salt Lake City: University of Utah Press.

Hochberg, M.C. 1980. Factors Affecting Leaf Size of Chaparral Shrubs on the California Islands. In *The California Islands: Proceedings of a Multidisciplinary Symposium*, ed. Dennis M. Power, 189–206. Santa Barbara Museum of Natural History: Santa Barbara.

Jazwa, Christopher S. 2015. A Dynamic Ecological Model for Human Settlement on California's Northern Channel Islands. Doctoral dissertation, Department of Anthropology, Pennsylvania State University, University Park.

Jazwa, Christopher S., and Rod Mather. 2014. Archaeological Site or Natural Marine Community? Excavation of a Submerged Shell Mound in Ninigret Pond, Rhode Island. *The Journal of Island and Coastal Archaeology* 9: 268–288.

Jazwa, Christopher S., and Jennifer E. Perry. 2013. Introduction. In *California's Channel Islands: The Archaeology of Human-Environment Interactions*, ed. Christopher S. Jazwa, and Jennifer E. Perry, 1–4. Salt Lake City: University of Utah Press.

Jazwa, Christopher S., Douglas J. Kennett, and Bruce Winterhalder. 2013. The Ideal Free Distribution and Settlement History at Old Ranch Canyon, Santa Rosa Island. In *California's Channel Islands: The Archaeology of Human-Environment Interactions*, ed. Christopher S. Jazwa, and Jennifer E. Perry, 75–96. Salt Lake City: University of Utah Press.

Jazwa, Christopher S., Todd J. Braje, Jon M. Erlandson, and Douglas J. Kennett. 2015a. Central Place Foraging and Shellfish Processing on California's Northern Channel Islands. *Journal of Anthropological Archaeology* 40: 33–47.

Jazwa, Christopher S., Douglas J. Kennett, and Bruce Winterhalder. 2015b. A Test of Ideal Free Distribution Predictions Using Targeted Survey and Excavation on California's Northern Channel Islands. *Journal of Archaeological Method and Theory*. doi:10.1007/s10816-015-9267-6.

Jazwa, Christopher S., Christopher J. Duffy, Lorne Leonard, and Douglas J. Kennett. 2016. Hydrological Modeling and Prehistoric Settlement on Santa Rosa Island, California. *Geoarchaeology: An International Journal* 31: 101–120.

Johnson, Donald L. 1980. Episodic Vegetation Stripping, Soil Erosion, and Landscape Modification in Prehistoric and Recent Historic Time, San Miguel Island, California. In *The California Islands: Proceedings of a Multidisciplinary Symposium*, ed. Dennis M. Power, 103–121. Santa Barbara Museum of Natural History: Santa Barbara.

Johnson, Donald L. 1989. Subsurface Stone Lines, Stone Zones, and Biomantles Produced by Bioturbation via Pocket Gophers (*Thomomys bottae*). *American Antiquity* 54: 370–389.

Johnson, John R. 1982. An Ethnographic Study of the Island Chumash. Masters thesis, Department of Anthropology, University of California, Santa Barbara.

Johnson, John R. 1993. Cruzeño Chumash Social Geography. In *Archaeology on the Northern Channel Islands of California*, ed. Michael A. Glassow, 19–46. Salinas, CA: Coyote Press.

Johnson, John R. 2001. Ethnohistoric Reflections of Cruzeño Chumash Society. In *The Origins of a Pacific Coast Chiefdom: The Chumash of the Channel Islands*, ed. Jeanne E. Arnold, 21–52. Salt Lake City: University of Utah Press.

Johnson, John R., Thomas W. Stafford, Jr., H.O. Aije, and Don P. Morris. 2002. Arlington Springs Revisited. In *The Fifth California Islands Symposium*, David R. Browne, Kathryn L. Mitchell, and Henry W. Chaney, eds., 541–545. Santa Barbara Museum of Natural History, Santa Barbara.

Jones, Terry L., and Al Schwitalla. 2008. Archaeological Perspectives on the Effects of Medieval Drought in Prehistoric California. *Quaternary International* 188: 41–58.

Jones, Terry L., Gary M. Brown, L. Mark Raab, Janet L. McVicar, W. Geoffrey Spaulding, Douglas J. Kennett, Andrew York, and Phillip L. Walker. 1999. Environmental Imperatives Reconsidered: Demographic Crises in Western North America during the Medieval Climatic Anomaly. *Current Anthropology* 40(2): 137–170.

Junak, Steve, Tina Ayers, Randy Scott, Dieter Wilken, and David Young. 1995. *A Flora of Santa Cruz Island*. Santa Barbara: Santa Barbara Botanic Garden.

Kennett, Douglas J. 1998. Behavioral Ecology and the Evolution of Hunter-Gatherer Societies on the Northern Channel Islands, California. Doctoral dissertation Department of Anthropology, University of California, Santa Barbara.

Kennett, Douglas J. 2005. *The Island Chumash, Behavioral Ecology of a Maritime Society*. Berkeley: University of California Press.

Kennett, Douglas J., James P. Kennett, Jon M. Erlandson, and Kevin G. Cannariato. 2007. Human Responses to Middle Holocene Climate Change on California's Channel Islands. *Quaternary Science Reviews* 26: 351–367.

Kennett, D.J., J.P. Kennett, G.J. West, J.M. Erlandson, J.R. Johnson, I.L. Hendy, A. West, B. J. Culleton, T.L. Jones, and Thomas W. Stafford Jr. 2008. Wildfire and Abrupt Ecosystem Disruption on California's Northern Channel Islands at the Ållerød-Younger Dryas Boundary (13.0–12.9 ka). *Quaternary Science Reviews* 27: 2528–2543.

Kennett, Douglas J., Bruce Winterhalder, Jacob Bartruff, and Jon M. Erlandson. 2009. An Ecological Model for the Emergence of Institutionalized Social Hierarchies on California's Northern Channel Islands. In *Pattern and Process in Cultural Evolution*, ed. Stephen Shennan, 297–314. Berkeley: University of California Press.

Muhs, Daniel R. 1987. Geomorphic Processes in the Pacific Coast and Mountain System of Central and Southern California. In *Geomorphic Systems of North America*, ed. William L. Graf, 560–570. Boulder, CO: Geological Society of America.

Perry, Jennifer E. 2003. Changes in Prehistoric Land and Resource Use among Complex Hunter-Gatherer-Fishers on Eastern Santa Cruz Island, California. Doctoral dissertation, Department of Anthropology, University of California, Santa Barbara.

Perry, Jennifer E. 2013. The Archaeology of Ritual on the Channel Islands. In *California's Channel Islands: The Archaeology of Human-Environment Interactions*, ed. Christopher S. Jazwa, and Jennifer E. Perry, 137–155. Salt Lake City: University of Utah Press.

Perry, Jennifer E., and Colleen Delaney-Rivera. 2011. Interactions and Interiors of the Coastal Chumash: Perspectives from Santa Cruz Island and the Oxnard Plain. *California Archaeology* 3: 103–126.

Raab, L.Mark, and Daniel O. Larson. 1997. Medieval Climatic Anomaly and Punctuated Cultural Evolution in Coastal Southern California. *American Antiquity* 62(2): 319–336.

Reeder-Myers, Leslie, Jon M. Erlandson, Daniel R. Muhs, and Torben C. Rick. 2015. Sea Level, Paleogeography, and Archaeology on California's Northern Channel Islands. *Quaternary Research* 83: 263–272.

Reitz, Elizabeth J., and Elizabeth S. Wing. 1999. *Zooarchaeology*. Cambridge: Cambridge University Press.

Rick, Torben C. 2009. 8000 Years of Human Settlement and Land Use in Old Ranch Canyon, Santa Rosa Island, California. In *Proceedings of the Seventh California Islands Symposium*, ed. Christine C. Damiani, and David K. Garcelon, 21–32. Institute for Wildlife Studies: Arcata, CA.

Rick, Torben C., Jon M. Erlandson, and René L. Vellanoweth. 2001. Paleocoastal Marine Fishing on the Pacific Coast of the Americas: Perspectives from Daisy Cave, California. *American Antiquity* 66(4): 595–613.

Rick, Torben C., Jon M. Erlandson, René L. Vellanoweth, and Todd J. Braje. 2005. From Pleistocene Mariners to Complex Hunter-Gatherers: The Archaeology of the California Channel Islands. *Journal of World Prehistory* 19: 169–228.

Rick, Torben C., Jon M. Erlandson, and René L. Vellanoweth. 2006. Taphonomy and Site Formation on California's Channel Islands. *Geoarchaeology: An International Journal* 21 (6):567–589.

Rick, Torben C., Jon M. Erlandson, Todd J. Braje, James A. Estes, Michael H. Graham, and René L. Vellanoweth. 2008. Historical Ecology and Human Impacts on Coastal Ecosystems of the Santa Barbara Channel Region, California. In *Human Impacts on Ancient Marine Ecosystems*, ed. Torben C. Rick, and Jon M. Erlandson, 77–101. Berkeley: University of California Press.

Rick, Torben C., Jon M. Erlandson, Nicholas P. Jew, and Leslie A. Reeder-Myers. 2013. Archaeological Survey, Paleogeography, and the Search for Late Pleistocene Paleocoastal Peoples of Santa Rosa Island, California. *Journal of Field Archaeology* 38(4): 321–328.

Schiffer, Michael B. 1987. *Formation Processes of the Archaeological Record.* Albuquerque: University of New Mexico Press.

Schoenherr, Allan, A., C. Robert Feldmeth, and Michael J. Emerson. 1999. *Natural History of the Islands of California.* Berkeley: University of California Press.

Stine, Scott. 1994. Extreme and Persistent Drought in California and Patagonia during Medaeval Time. *Nature* 369: 546–549.

Stright, Melanie J. 1990. Archaeological Sites on the North American Continental Shelf. In *Archaeological Geology of North America*, ed. Norman P. Lasca, and Jack Donahue, 439–465. Boulder, CO: Geological Society of America.

Timbrook, Jan. 1993. Island Chumash Ethnobotany. In *Archaeology on the Northern Channel Islands of California*, Michael A. Glassow, ed., 47–62. Coyote Press, Salinas, CA.

Timbrook, Jan. 2007. *Chumash Ethnobotany: Plant Knowledge among the Chumash People of Southern California.* Santa Barbara Museum of Natural History, Santa Barbara.

Watts, Jack, Brian Fulfrost, and Jon M. Erlandson. 2011. Searching for Santarosae: Surveying Submerged Landscapes for Evidence of Paleocoastal Habitation Off California's Northern Channel Islands. In *The Archaeology of Maritime Landscapes*, ed. Ben Ford, 11–26. New York: Springer.

Winterhalder, Bruce, Douglas J. Kennett, Mark N. Grote, and Jacob Bartruff. 2010. Ideal Free Settlement on California's Northern Channel Islands. *Journal of Anthropological Archaeology* 29: 469–490.

Wood, W. Raymond, and Donald L. Johnson. 1978. A Survey of Disturbance Processes in Archaeological Site Formation. *Advances in Archaeological Method and Theory* 1:315–381.

Yatsko, Andrew. 2000. From Sheepherders to Cruise Missiles: A Short History of Archaeological Research at San Clemente Island. *Pacific Coast Archaeological Society Quarterly* 36(1): 18–24.

Chapter 9
Formation Processes of Maritime Archaeological Sites in Guadeloupe (French West Indies): A First Approach

Jean-Sébastien Guibert, Christian Stouvenot and Frédéric Leroy

Introduction

In his 1804 description of Morel Beach on the Caribbean island of Guadeloupe, French colonial governor, General Jean Augustin Ernouf, relates how he visited "... this famous coast where are found human fossils wrapped in petrified madrepores [a genus of stony coral]" (Faujas-Saint-Fond 1804; Delpuech 2005). The formation of beachrock, a fast-cementing sandy rock, was poorly understood in the early nineteenth century. The "fossil-like" Amerindian burials encased in beachrock were at that time mistaken by many scholars as testimony to the great antiquity of man in the Caribbean, and by extension in the world.

J.-S. Guibert (✉)
AIHP-GÉODE EA 929 Université des Antilles,
UFR Lettres et Sciences Humaines Campus de Schoelcher,
BP 7207 97275, Schoelcher Cedex, France
e-mail: jean-sebastien.guibert@univ-antilles.fr

C. Stouvenot
ArchAm (Archéologie des Amériques), Maison Archéologie & Ethnologie
René-Ginouvès, CNRS UMR 8096, 21, Allée de L'Université,
92023 Nanterre Cedex, France
e-mail: christian.stouvenot@culture.gouv.fr

C. Stouvenot
Service Régional de L'Archéologie de la Guadeloupe 28,
Rue Perrinon, 97100 Basse-Terre, Guadeloupe, France

F. Leroy
Département des Recherches Archéologiques Subaquatiques et Sous-Marines
Direction Générale des Patrimoines, Ministère de la Culture et de la Communication,
UMR 5140 (Archéologie des Sociétés Méditerranéennes), CNRS—UPV—MCC—INRAP
147 Plage de l'Estaque, 13016 Marseille, France
e-mail: frederic.leroy@culture.gouv.fr

© Springer International Publishing AG 2017
A. Caporaso (ed.), *Formation Processes of Maritime Archaeological Landscapes*,
When the Land Meets the Sea, DOI 10.1007/978-3-319-48787-8_9

This remarkable coastal discovery illustrates a unique and fascinating form of archaeological site formation process in maritime tropical environments. Archaeological research in the West Indies has not focused specifically on this process; however, interest in coastal formation processes is growing. This is evident in the multidisciplinary design of many current and ongoing archaeological projects. Recent examples in Guadeloupe include: "The Caille-à-Bélasse Project", a pre-Columbian coastal shell midden on Petite-Terre island (Gagnepain 2007, 2010); "The Entre terre et mer Project", a regional land survey of coastal colonial buildings and structures around Petit and Grand Cul-de-Sac-Marin (Guibert and Bigot 2014); and "The D'îles en îles Project," a new investigation of two micro-islets near Guadeloupe and their surroundings by underwater archaeological methods and technology (Guibert et al. 2013). In Saint Martin, numerous archaeological salvage excavations have provided unprecedented quantities of data about multicomponent preceramic site formation, controlled by storm activity and the sites' geomorphological contexts (Bonnissent 2008; Bonnissent et al. 2013). On the same island, studies, based on test excavation, demonstrate the dependency status between several remote pre-Columbian coastal and micro-island sites (Bonnissent 2008). In Martinique, between 2009 and 2013, a Venezuelan team scouted archaeological sites on several little islands off the Atlantic coast using terrestrial pedestrian surveys and test excavation (Antczak et al. 2012, 2015). Understanding and taking into account these formation processes is also providing new insight in the reanalysis of historic accounts and previous archaeological efforts.

The approach to archaeological site and landscape formation processes proposed in this chapter considers the entirety of the northern French West Indies: Guadeloupe; Saint Barthélemy; and French Saint Martin (Fig. 9.1), from the Pre-Columbian (3000 BC–1500 AD) through the colonial period (1500 AD–1900 AD). The inclusion of Amerindian sites allows for the development of a theoretical approach to landscape formation that combines long-term natural and anthropogenic formation processes with relatively short-term (i.e., no more than 500 years) formation processes acting on colonial period sites. Maritime archaeological sites are defined as submerged and coastal terrestrial sites whose natural formation transforms are primarily controlled by marine processes, as in the example of the Amerindian burials encased in beachrock. Archaeological sites covered in this analysis include Amerindian isolated finds in addition to multicomponent sites, whereas only sites with extant structural components, including shipwrecks, are considered for the colonial period.

In the coastal context of the French West Indies, the sea may be considered as a primary control in overall landscape and individual site formation processes, but its role as a control is poorly understood, except in the preceramic contexts of Saint-Martin (Bonnissent et al. 2013). This chapter aims to present a series of "potentialities" of landscape and site formation processes active within this region using several recently analyzed archaeological examples.

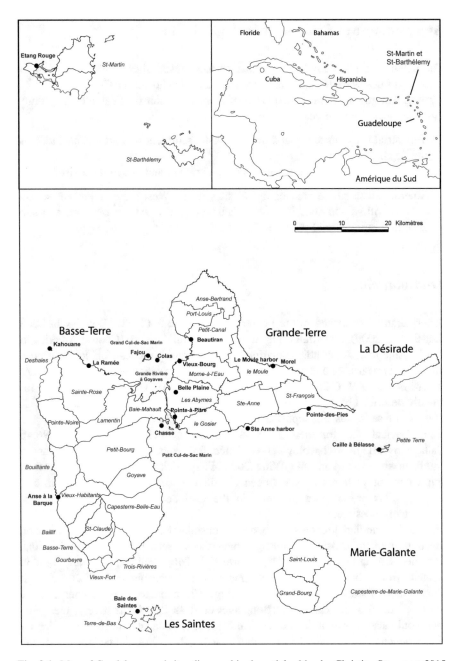

Fig. 9.1 Map of Guadeloupe and sites discussed in the article. *Map* by Christian Stouvenot 2015

Amerindian Maritime Archaeological Sites

The potential for Amerindian maritime archaeological sites in the islands of the Lesser Antilles was first described by Desmond Nicholson (1976). Based on his work in Antigua, he defined two types of maritime archaeological sites in addition to those located on the coast

1. Terrestrial sites located inland due to coastal sediment accretion and/or tectonic uplift; and
2. Submerged sites inundated due to coastal erosion and/or sea level rise.

Amerindian maritime archaeological sites in the West Indies represent several types of maritime activities including transportation; coastal habitation; and subsistence and raw material procurement.

Transportation

Amerindians first arrived in the Lesser Antilles circa 3000 BC (Bonnissent 2008; Siegel et al. 2009), potentially migrating from South America, Central America, or North America. Amerindian mariners used large pirogues for long-distance, open-sea voyages, and smaller vessels for coastal activities (Nicholson 1977; Bérard et al. 2009; Callaghan 2011; Fitzpatrick 2013). These practices were frequently described by contemporary chroniclers (Breton 1656:69; Fitzpatrick 2013). The use of sails was not reported by European explorers, and the historical and archaeological data for their potential use are inconclusive (Gannier 1996:46; Callaghan 2011). Archaeological evidence for canoes and paddles is rare at Antillean sites (Newsom and Wing 2004; Fitzpatrick 2013). No pre-Columbian shipwrecks of such vessels have been identified in the Lesser Antilles, but, considering their preservation potential in the Caribbean, their discovery remains theoretically possible.

The Amerindian logboats were nearly unsinkable. Contemporary chroniclers mention frequent capsizings during which the vessels, passengers, and ladings remained safe and undamaged (De la Porte 1775:246). Ladings, however, could be ejected from the vessel. Recent underwater archaeological discoveries in Guadeloupe of stone pestles, millstones (Fig. 9.2), and axes may be interpreted as the result of pirogues capsizing (Boulanger et al. 2012). Alternatively, these artifacts could also represent the scattered remains of an inundated terrestrial settlement site. Stouvenot is currently directing the "Macou 2015, Guadeloupe" project on this topic.

Fig. 9.2 Pre-columbian grindstone found at 4 m depth near Macou Islet, Grand-Cul-de-Sac-Marin. *Photo* by Jean-Sébastien Guibert 2012

Coastal Habitation

Amerindian settlements are frequently located in a coastal context (Fig. 9.3). Habitation sites include villages, temporary encampments, and shell middens. Contemporary chroniclers report that Late Amerindian village sites were characterized by wooden huts called "carbets" (Breton 1665) with peripheral middens as shown by archaeological excavations (Rodriguez 1992; Ramcharan 2004:162–163; Bonnissent 2008:128, 203). Since the 1960s, archaeological research in Guadeloupe and the northern French West Indies has identified nearly 40 major terrestrial village sites adjacent to the coast, including cave settlements and rock art sites (DAC Guadeloupe 2015).

Stilt houses, such as the well-known Cuban lagoon site of Los Buchillones, where the house remains have collapsed into the sea (Graham et al. 2000) have not been identified in Guadeloupe. Similar sites may exist in the Lesser Antilles, most likely in association with large mangrove or coastal swamps as in Grand Cul-de-Sac-Marin in Guadeloupe. One potential site has been identified in Guadeloupe; in the city of Saint François, several worked wooded posts have been recovered in a swamp nearby a Saladoid habitation site (Martias 2005).

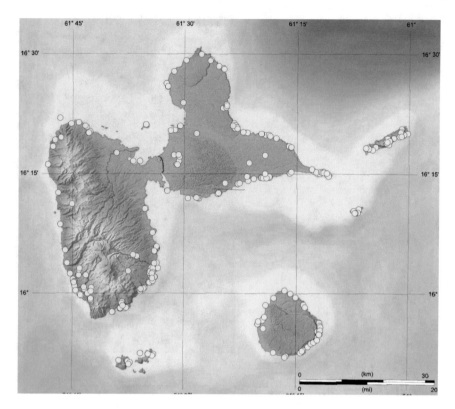

Fig. 9.3 Pre-columbian sites in Guadeloupe. National Archaeological map, DAC Guadeloupe 2015/Fond topographique Guadeloupe: Eric Gaba—Wikimedia Commons

Cave settlements and rock art sites of the Lesser Antilles are interpreted as ceremonial, mortuary, or refuge sites (Grouard et al. 2013). They are regularly located near the sea or at the mouths of rivers. As such, they may be considered oriented toward maritimity.

Food and Material Procurement

Shellfish gathering was one of the most common and widespread Amerindian coastal procurement activities. The shucking of shellfish over long time periods resulted in the creation of shell mounds, a frequent archaeological site type encountered along the coast. One of the most important shell mound sites is called the "Lambi Line," located on Fajou Island (Grand Cul-de-Sac Marin, Guadeloupe).

The Lambi Line extends for 900 m in length (Stouvenot 2006). Another large shell mound is identified on the Kahouanne Islet; it measures approximately 130 m length and several meters high (Yvon 2011). Semi-submerged shell mounds are unknown in Guadeloupe; however, several have been identified in Anegada, US Virgin Islands (IRF 2013:131). These types of mounds began their formation by anthropogenic discard of conch shells from a boat in shallow waters. The process continued until the mound emerged from the water.

In general, micro-island sites may be interpreted as specialized hunting, fishing, and shellfish gathering locations (Gagnepain et al. 2007). This site type and location almost exclusively dates to the Late Ceramic Age (Late Neoindian period), after approximately 1000 AD (Bérard et al. 2005). Recent underwater investigations have been undertaken in the vicinity of two micro-islands: Chasse Island and Colas Island, located in Petit and Grand Cul-de-Sac Marin, respectively, confirming this site type identity and age.

The Natural Context of Formation, Transformation, or Destruction Processes of Amerindian Sites

Three types of long-term, natural formation process regimes may be proposed for Amerindian archaeological sites

Type A: Habitation sites that are preserved in a strictly coastal context

Most Type A sites are well preserved, but some at low elevation are suffering storm and tidal erosion resulting in potentially significant site loss. The majority of the village sites cited in this chapter are representative of Type A.

Type B: Habitation and resource procurement sites formed initially on the coast that are now located and preserved inland due to sediment accretion and/or tectonic or co-seismic uplift

The Strombus Line of Barbuda, dating to circa 2000 BC, is a shell midden, once located along the shoreline that now stretches for more than 4 km across the island (Watters and Donahue 1990; Watters 2001). In Guadeloupe, the Neoindian village Belle Plaine, dating to circa 1200 AD and now located 2 km inland (Fig. 9.4), contains a shell midden suggesting previous proximity to the sea. Potentially initially formed as a coastal site, Belle Plaine is now separated from the sea by a poorly accessible mangrove and fresh water swampy forest (Stouvenot 2010).

Type C: Sites that are partially or completely submerged due to relative sea level rise caused by post-glacial transgression and/or local tectonic or co-seismic movements

There is little conclusive evidence for this site type in the Lesser Antilles. Circa 3000 BC, the sea level may have been as much as 3–6 m lower than today (Fig. 9.5) (Toscano and Macintyre 2003). Two archaeological sites in the French

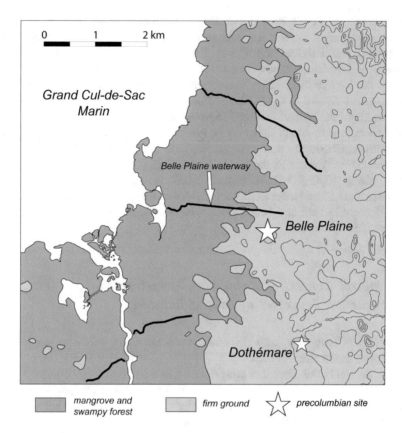

Fig. 9.4 Map showing location of the Belle Plaine Pre-columbian site. *Map* by Christian Stouvenot 2015

West Indies provide potential evidence for this site type. Étang-Rouge in Saint Martin, dating to circa 3300 BC, is interpreted as a temporary campsite associated with shellfish gathering (Bonnissent 2008). The site is buried under a 6 m thick sandbar. Pointe des Pies site in Saint-François, Guadeloupe, dating to circa 400 BC, is a preceramic site where archaeological remains are lying in mangrove peat 20 cm below current sea level (Richard 1994). The scarcity of identified preceramic sites may be linked to sea level rise, the effects of which may have submerged or destroyed associated archaeological remains.

In the case of potential submerged micro-island sites, it remains to be evaluated whether the cause of site submergence is due to coastal site erosion, sea level rise, or by tectonic or co-seismic activity.

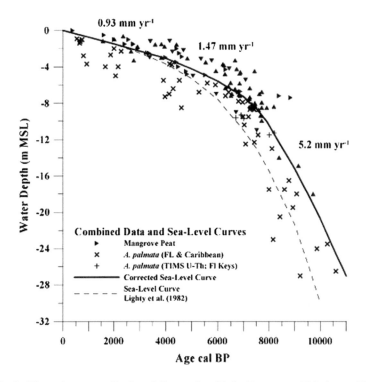

Fig. 9.5 Caribbean Quaternary Sea Level Curve. *Graphic* by Toscano and Macintyre 2003

Colonial Maritime Archaeological Sites

Coastal Colonial Sites

Colonial maritime archaeological sites are those primarily associated with colonial maritime activities. For the colonial period, two types of maritime archaeological sites are defined: sites associated with plantation activities and ports.

Plantation sites may be on the coast; however, most associated structures are located well away from the seashore (Bégot 1991). Only the purposefully coastal structures of plantation sites, such as wharves or quays, are adjacent to the sea. Wharf and quay sites have been systematically studied in Grand Cul-de-Sac Marin. An analysis of this colonial maritime landscape reveals the evolution of eighteenth to mid-twentieth century structures (Guibert and Bigot 2014). For example, a wharf associated with the late eighteenth century La Ramée plantation is comprised of a stone pier, with continuous formwork made from Pink Trumpet wood (*Tabebuia aff. Heterophylla*). The structure dates to the late eighteenth to middle nineteenth centuries.

On the other bank of Grand Cul-de-Sac Marin, the mid-nineteenth century site of Beautiran contains quite different quay structures. Three transhipment zones have been identified. To the north, the eighteenth century transhipment zone is located on the natural embankment. One dating to the second half of the nineteenth century linked to the Beauport factory, which is composed of a right angle dock, a loading gantry, and a railroad access (Barfleur 2005). Lastly, one dating to the twentieth century to the south includes a ramp and a cradle constructed of cement and concrete.

These different structures illustrate the evolution of local transportation regimes. They represent the main phases of colonial sugar development, from the sugar plantation to the central factory. These main phases modified the littoral landscape. After a period of occasional use of multiple wharves, canals, and embankments in the eighteenth and nineteenth centuries, transhipment activity was concentrated at warehouse or redistribution centers at the end of the nineteenth and the early twentieth centuries. Today, those located in towns have been transformed into fishing ports or marinas while those that are isolated are left abandoned, in some cases anticipating heritage protection and improvements.

The importance of coastal plantation sites in the maritime landscape is evident; structures, such as canals associated with maritime activity continue to demarcate the territory today. In the seventeenth century, their use was associated with colonization and plantation development, but they also progressively fixed the presence of villages such as Petit-Canal, Vieux-Bourg, and even Morne-à-l'Eau in the beginning of the nineteenth century. In several cases, these structures permitted the development of specific activities such as pottery in Vieux-Bourg and maritime trade and fishing in Petit-Canal (Guibert and Bigot 2014). Today their use is associated with a fishery and tourism; however, their historic importance is not easily perceptible in the modern landscape.

This is not the case at Beautiran where both quays and vessel remains are visible in both the terrestrial and littoral contexts. The site was the local port of the Beauport Sugar Factory (1863–1990). The primary factory site is located inland and has been restored and is open to the public, but the coastal component, Beautiran, is abandoned. Docks, storage buildings, and workshops are still standing. A tourism development project at Beautiran is on hold because of the site's archaeological potential (Barfleur 2005). At least eight shipwrecks dating from circa 1900 to the 1960s are located either in the water or in the workshop area of the site. The site is isolated and difficult to access. Its conservation would preserve a unique twentieth century component of the archaeological landscape and provide key information in the analysis of the formation of the landscape into modern time.

Port Royal, Jamaica (Hamilton 2008), English Harbour, Antigua (Murphy 2003), and Saint-Pierre, Martinique (Serra 2012) all reveal the potential and the limitations of excavation in primary ports in the West Indies. These places are all together a record of maritime activity, commercial networks, and maritime risk through the presence of shipwrecks, coastal structures, and material culture accumulated in port middens.

No underwater archaeological projects have taken place in the main ports of Guadeloupe; however, Basse-Terre and Pointe-à-Pitre represent great potential for underwater archaeology. In spite of several harbor development projects, except for a short investigation of a shipwreck located within the harbor entrance (see below), no archaeological research has occurred at Pointe-à-Pitre harbor. Several oral accounts attest to the destruction of a shipwreck site at the harbor in the 1990s. This lack of cultural resource management represents a real threat to cultural heritage preservation and a concomitant loss of data that can reveal how the maritime archaeological landscape has formed.

The secondary ports of Petit-Canal and Le Moule have been partially investigated. Petit-Canal contains both structures and artifacts dating from the eighteenth century indicating the vitality of secondary ports (Bigot and Vicens 2006; Bigot and Guibert 2011a, b). Le Moule contains mid-nineteenth century structures and anchors on a coral reef placed to facilitate mooring. Fifty-four moorings, including 22 anchor moorings were recorded on both sides of the anchorage (Texier and Casagrande 2011). The presence of the latter is quite unusual. It is clear that the anchors were placed to mitigate the risk of mooring in this harbor, which was often subject to strong swell. These anchors might also have been used to facilitate maneuvers to both enter and exit the harbor (Fig. 9.6). During the nineteenth century at Le Moule, vessels would exchange cargoes of coal and machinery for sugar and rum. This activity declined by the end of the nineteenth century as the maritime activity at Le Moule changed following the sugar production crisis. These anchors illustrate the importance of this maritime activity, as well as represent its associated physical risk and behavioral response on the landscape. Today, this maritime landscape is put in perspective with the port activity and transatlantic role of Le Moule in the nineteenth century contrasted with its role today as a small fishing port.

Submerged Shipwreck Sites

Approximately 20 historic shipwreck sites have been inventoried in Guadeloupe (Guibert 2013); (Fig. 9.7). This analysis focuses only on wooden-hulled

Fig. 9.6 Le Moule harbor anchors. *Photo* by Pierre Texier 2012

shipwrecks, and excludes those constructed of iron. In 1999, the first archaeological excavations of shipwrecks in Guadeloupe took place in Anse à la Barque (Guibert 2010a). Three of the eight identified shipwrecks have been the subject of archaeological survey. Other recent investigations include the previously mentioned shipwreck at the entrance of Pointe-à-Pitre (Guibert and Bigot 2014) and a shipwreck at the entrance to Sainte-Anne (Guibert 2014). Research in progress may allow for the identification of one or more of the vessels in Les Saintes.

Stratigraphic interpretation of these sites made during archaeological testing appears to be supported by the approximately 15 other sites not yet investigated, but which have undergone reconnaissance survey. For deep water, low profile sites, sedimentation controls site formation processes. The cargo and/or ballast are, in several cases, a paramount factor in relative hull preservation. Shipwrecks AB1 and AB2 (Anse à la Barque) are located at a depth of 4–5 m in alluvium. Both contain a large amount of stone and/or cannon ballast. AB1 is even stuck under two others wrecks called AB4 and AB5, which are as yet unstudied.

The PAP1 (Pointe-à-Pitre) shipwreck site is embedded within a coral reef, located at a depth of 4–5 m. PAP1 may be the wreck of a merchant vessel known as *France* lost in 1824. A small portion of the wooden hull, three frames, is extant due to a protective covering of its stone cargo and two cannon. The weight of the stone cargo has crushed the exposed wooden remains of the vessel and teredo worm and wave energy has contributed significantly to site degradation and dispersal (Guibert 2014).

The PSA1 (Sainte-Anne) shipwreck is a relatively well-preserved, copper-sheathed wooden vessel, located at a depth of 7–8 m. Extant hull remains

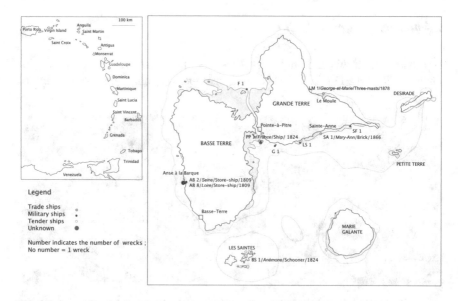

Fig. 9.7 Location of colonial shipwrecks. *Map* by Jean-Sébastien Guibert 2013

Fig. 9.8 The *Saint-Anne* shipwreck site. *Photo* by Jean-Sébastien Guibert 2014

include approximately 25 m of the portion of the vessel that would have been below the waterline, in this case the keel and the lower part of the frames (there is no evidence for the preservation of the upper part of the frames, keelson or upper works) (Guibert 2014). Visually striking bronze bolts, however, do remain in situ (Fig. 9.8). Of particular interest is evidence that the keel has partially decomposed; test excavation has not revealed the presence of a counter keel. The disposition of the extant remains of PSA1 illustrates the control that the composition of the seafloor plays in site formation processes.

The depth of the shipwreck may be decisive in different cases as in the case of the Moule site or Saintes site. The shipwreck at Le Moule, LM1, lies at 18 m depth, and the open-sea fetch is balanced by the fact that the wreck is stuck between two reefs. Despite only having reconnaissance survey, the good state of preservation of this site allows for a potential interpretation of it to be the wreckage of *George-et-Marie*, a French three-masted vessel lost in 1878 (Guibert 2014). The shipwreck in La Baie des Saintes, BS1, is also located at a relatively significant depth of 25 m. It could be the wreckage of *Anémone*, a schooner the French Restauration navy lost in 1824 (Guibert 2014). *Anémone* underwent archaeological investigation in July 2015, and it seems clear that the depth of the site is a favorable factor in its conservation (Guibert 2014).

The assessment of site disposition must be contextualized with an analysis of maritime behavior, the proximal cause of shipwreck site formation. Archival research has revealed records of approximately 550 historic maritime accidents: 78% due to weather conditions; 7.5% due to naval engagements; and 3.8% due to human error (Guibert 2013). A critical reading of historical sources permits the interpretation that only between 49 and 125 of these accidents may have resulted in presently extant archaeological sites. Pointe-à-Pitre well demonstrates this disparity. This area, where accidents were known to be numerous because of its high traffic and hazardous entrance, is also where archaeological potential is low because of ease of salvage. In the last 50 years, only four shipwreck sites have been identified, three in the entrance channel. The ratio of reported accidents to identified archaeological sites is 0.19 (Guibert 2010b).

The visibility of the submerged archaeological landscape in Guadeloupe is low. Except for the substantial late-nineteenth century shipwrecks at Le Moule and Sainte-Anne, most submerged archaeological remains are invisible to the untrained eye. Moreover, the composition of the seafloor and the structure of benthic communities may mask the presence of ballast piles and extant hull components.

Amerindian and Colonial Site Formation Processes

Environmental controls governing formation processes of colonial period archaeological sites may be similar to those of Amerindian ones, but may be less visible because of the smaller amplitude of the clues due to the shorter time evolution. It is difficult to link an individual site with a single systemic formation process pathway. The following presents interpretations of potential complex and simultaneous formation process pathways identified through multidisciplinary research.

Geological data indicates that post-glacial sea level rise would have significantly impacted the shoreline only until approximately 1000 AD, when the sea level became approximatively the same as today (Toscano and Macintyre 2003). However, Amerindian remains identified off micro-islands radiocarbon date to 1150 AD for Colas Island and 1350 AD for Chasse Island (Guibert et al. 2013), and their low position (0.5–1.5 m below actual sea level) is consequently not due to sea level rise. The geological observations throughout the Caribbean indicate that around 5000 years ago the sea level may have been approximately 3–6 m lower than today (Toscano and Macintyre 2003). Therefore at that time, Grand Cul-de-Sac Marin was a wide plain almost completely subaerial (Fig. 9.9). This paleolandscape would have been very suitable for human life and adaptation. Some submarine Amerindian remains have been recently identified in this region, but have yet to be dated or analyzed.

Tectonic subsidence has also been proposed for the formation of the low region of Grand Cul-de-Sac Marin (Garrabé and Andréieff 1988); however, this subsidence magnitude is not well established. At Les Saintes, Guadeloupe, it is hypothesized to be about 40 cm per thousand years (LeClerc et al. 2014), which is very significant at archaeological time scales.

Another possible hypothesis to explain the submergence of micro-island sites is the "elastic bounce" phenomenon, the faster part of co-seismic movements. It

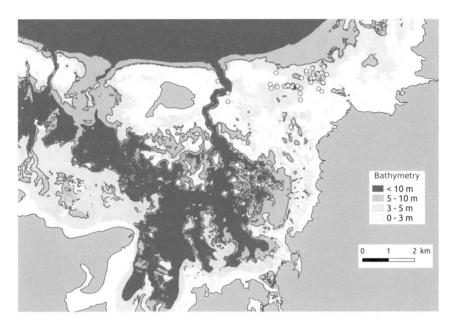

Fig. 9.9 Grand-Cul-de-Sac-Marin Bathymetry: *red line* = −5 m depth (around 3000–5000 BC), *yellow dots* isolated lithic artifacts. *Map* by Christian Stouvenot 2015, DAC Guadeloupe, with QGIS and Litto3D©IGN

results in sudden and permanent vertical shifts in the Earth's crust. It is the consequence of subduction zone slip and resultant earthquakes (Chlieh 2003). The occurrence of such movements has been observed in the Lesser Antilles by geophysicists, but their potential maximum amplitude is not yet determined (Weil Accardo et al. 2010).

Changes in the distribution of marine mass sediments can also significantly affect the stability of the micro-islands. These changes could be caused by human activity in the immediate vicinity or even much further away, in the watershed of rivers. For example during the nineteenth century, chalk or coal production produced on micro-islands required large amount of wood resulting in deforestation, which may have had an impact on shoreline erosion (Yvon 2012). It has also been demonstrated that the development of agriculture during the colonial period has altered the sedimentary output of local rivers. For example a dramatic increase of sedimentary accretion has been observed in Grand Cul-de-Sac Marin at the mouth of the Grande Rivière à Goyave (Ingénieurs du Roi 1768) (Fig. 9.10). More recently, port construction projects, including dredging and shoreline modification, have had a substantial impact on beaches and coral reefs.

Erosion, due to successive storm and wave activity, is evident at several archaeological sites in Guadeloupe. In fact, violent tropical storms and tsunami waves are a primary controlling factor in coastal site formation processes. For example, 1821, 1824, or 1825 hurricanes appear to have significantly modified the

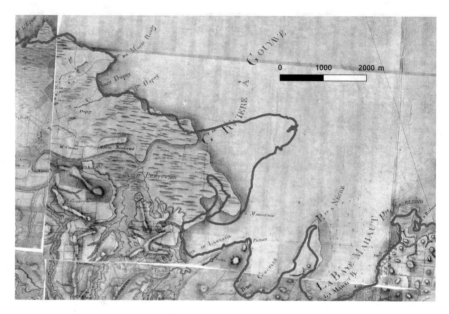

Fig. 9.10 Map of "Ingénieurs du Roi, 1768" in the region of the mouth of the Grande Rivière à Goyaves in 1768, and the location of the current coastline (*red line*). This image shows the significant progradation of the coastline (2 km) since 1768. (Map by Christian Stouvenot 2015 DAC Guadeloupe, with QGIS and Litto3D©IGN)

structure of Colas Island (Guibert et al. 2013). Its 1829 description is very different from that of 1818; the former describes a much smaller, uninhabited site. Smaller storms may continuously, but superficially impact surface sands; however, coastal building and damage to vegetation and coral reef structure is accelerating and is a primary cause of modern littoral erosion (Fitzpatrick 2010).

Storm and wave activity may explain the formation of the colonial period seawall site at Chasse Island (Guibert et al. 2013). The remains of a wood stake and stone structure are located between 25 and 30 m from the modern seashore at 60 cm depth.

Violent storms are likely a proximal cause of many coastal shipwrecks in the Lesser Antilles. It is difficult though to interpret the effect of storm and wave activity on submerged shipwreck sites due to a paucity of data. Marine biological studies have shown that corals found in proximity to the trajectories of storms, and at a depth of 0–15 m, are the most affected (Bouchon et al. 1991). It follows that for shipwrecks, depth, the sedimentation regime, and storm frequency are in all likelihood the causes of long-term site destruction. Storm activity can regularly uncover shipwrecks, and therefore it follows that it contributes to their destruction. Several shipwrecks at l'Anse à la Barque were discovered after individual storms.

Most of the shipwreck sites at Guadeloupe date from the beginning of the nineteenth century. Research on seventeenth century shipwrecks, such as that by Jean-Pierre Moreau (1992) on the Spanish galleons of the flotilla of Marquis de

Monteclaros destroyed on the coast of Basse-Terre, has proved negative. Eighteenth century shipwreck sites have been occasionally identified, but at this time, the oldest identified are the corvettes de charge (fluyts) *Seine* and *Loire* destroyed in 1809 (Guibert 2010a). Older sites have been located, but have not yet been investigated, for example off Gosier, the IG1 site is supposed to be from the eighteenth century, and is composed of a ballast mound and an isolated cannon

Conclusion

This chapter presents a pilot approach dealing with the question of formation processes in the Lesser Antilles, using Guadeloupe as an example. Different case studies show the potential risks to maritime and underwater archaeological sites in Guadeloupe, which need to be addressed if we are to record these sites before they are lost completely.

Processes of the formation and destruction of coastal and underwater archaeological sites can be classified into two categories: those resulting from slow geological processes acting in the long term, such as post-glacial isostatic rebound and local tectonic or co-seismic movement; and those resulting from sudden events, such as storms, tsunamis, and co-seismic activity. Long-term processes controlled by natural phenomena are fairly well identified for the pre-Columbian period because we can identify significant deposits that represent long periods of time that, at some sites, extend to considerable depths in the sedimentary record. Conversely, brief and abrupt events produce impacts more readable for more recent periods, the aging of the coast having the tendency to progressively erase traces of former events. Sudden events such as storms or earthquakes are also better documented for the colonial period, thanks to the presence of archival records. Nevertheless, this information, although valuable concerning the impact to society, generally lack direct and reliable observations of the impact of sudden events on coastal colonial archaeological sites or even on the precise way in which vessels wrecked.

In theory, the interaction between the land and sea can happen in two ways: without movement of material (submersion or emersion); or with movement of material (erosion or sedimentary accretion). In practice, we more often observe these slow/rapid events: submersion/emersion, erosion, sedimentary accretion, coupled in complex ways, with spatial and temporal variability in form and degree. In the Lesser Antilles, for several millennia, the trend is rather toward submersion and erosion although the reverse may occur locally and for short periods.

Low and flat coastlines and micro-islands represent for archaeologists and paleoenvironmentalists a sort of very interesting "laboratory" for the observation of these phenomena because the effects of vertical variations are found amplified horizontally due to the topography of this landscape. Land loss is evident from the pre-Columbian period and continues quite impressively through the historical period, leaving us to fear for the imminent disappearance of these sites due to the predicted rise in sea level as a result of global warming, also the acceleration in erosive processes related to the impact of uncontrolled land development, particularly along the coast.

This trend toward a loss of land area of course is worrying as regards the preservation of the quality of the natural and living environment. It severely impacts the integrity of coastal archaeological sites, such as, for example, several slave cemeteries in Guadeloupe where accelerated erosion has recently affected the part of the population particularly sensitized to this issue. Proposed mitigation is twofold: salvage excavation or stabilization measures to arrest erosion. The situation is taken very seriously by archaeologists because these measures are for the most part impractical due to their need for considerable manpower for salvage excavation, or due to the exorbitant costs of erosion control. Several Antillean territories find themselves confronted with these issues, and, after realization of the need for mitigation, archaeologists are now trying to cooperate, to share their experiences and ideas in their search for practical solutions to these problems, and also to evaluate the scale and the urgency of this threat. Therefore, in July 2015, during the colloquium of the International Association for Caribbean Archaeology, a round table entitled "Threatened Coastal Sites in the Caribbean" was organized around the theme of erosion of archaeological sites. It is the hope that these efforts will succeed in alerting land managers of the gravity of the problem, because this question cannot be solved by archaeologists alone; at stake is the preservation of unique and nonrenewable knowledge that is only provided by archaeological sites.

Acknowledgments This chapter would not have been possible without the help and support of Françoise Pagney, French West Indies University, Archaeology History Patrimony AIHP Geode director and geography professor; Danielle Bégot, French West Indies University, AIHP Geode history professor; Michel L'Hour, DRASSM Director. Special thanks to Alicia Caporaso for scientific exchanges and help in translation.

References

Antczak, Andrzej, Ma. Magdalena Mackowiak de Antczak, Konrad Antczak, and Olivier Antczak 2012. *Prospection archéologique sur l'îlet Madame en Martinique*. Caracas, Venezuela.

Antczak, A., Ma. Magdalena Antczak, Sébastien Perrot-Minnot, Konrad A. Antczak, and Olivier A. Antczak. 2015. *Rapport de la prospection archéologique des îlets Frégate, Long, Métrente et Thierry*. Fort-de-France, Martinique, France: Martinique.

Barfleur, Jean. 2005. *Sucre et mangrove Beautiran un port intérieur guadeloupéen*. France: Agence Warichi Conseil Général de Guadeloupe.

Bégot, Danielle 1991. *Les habitations-sucreries du littoral guadeloupéen et leur évolution*. pp. 151–190, Fort de France, Martinique, France: Caribena, Centre d'études et de recherches archéologiques de la Martinique.

Bérard, Benoît, Sandrine Grouard, and Nathalie Serrand 2005. L'occupation post-saladoïde du sud de la Martinique, une approche de l'idée de territoire. In *Proceedings of the 21st Congress of the International Association for Caribbean Archaeology*, Port of Spain, Trinidad.

Bérard, Benoit, Jean-Yves Billard, and Bruno Ramstein 2009. Ioumoulicou. In *Proceedings of the 23th Congress of the International Association for Caribbean Archaeology*, Antigua and

Barbuda, pp. 577–589. https://hal-uag.archives-ouvertes.fr/hal-00966520. Accessed Jun 15, 2015.

Bigot, Franck, and Jean-Sébastien Guibert. 2011a. *Entre terre et mer: Rapport de prospection.* AAPA, Petit-Bourg, Guadeloupe, France: Infrastructures littorales dans le Grand Cul-de-Sac Marin de la Guadeloupe.

Bigot, Franck, and Jean-Sébastien Guibert. 2011b. *Entre terre et mer: Rapport de prospections sondages sous-marins, Infrastructures littorales dans le Grand Cul de Sac Marin de la Guadeloupe.* AAPA, Petit-Bourg, Guadeloupe, France.

Bigot Franck and Bernard Vicens. 2006. *Le Port du Moule Rapport de prospection sondage.* Pointe-à-Pitre, Guadeloupe, France: CERC.

Boulanger, Marc, Jean-Sébastien Guibert, François Nouailhas, and Christian Stouvenot. 2012. *Découverte archéologique fortuite. Site de "Macou A". Guadeloupe. Rapport de prélèvement.* AAPA, Petit-Bourg, Guadeloupe, France.

Bonnissent, Dominique, 2008. *Archéologie précolombienne de l'île de Saint-Martin, Petites Antilles (3300 BC–1600 AD).* Université Aix-Marseille I—Université de Provence, Aix-en-Provence, France. http://tel.archives-ouvertes.fr/tel-00403026/fr/. Accessed Jun 15, 2015.

Bonnissent, Dominique, Pascal Bertran, Antoine Chancerel, Pierrick Fouéré, Sandrine Grouard, Franck Mazéas, Auran Randrianasolo, Thomas Romon, Nathalie Serrand, and Christian Stouvenot 2013. *Les gisements précolombiens de la Baie Orientale. Campements du Mésoindien et du Néoindien sur l'île Saint-Martin (Petites Antilles).* Editions de la maison des sciences de l'Homme, Documents d'archéologie française, Paris, France.

Bouchon, Claude, Yolande Bouchon Navaro, Daniel Imbert, Max Louis. 1991. L'impact de l'ouragan Hugo sur les écosystèmes côtiers de la Guadeloupe. *L'ouragan Hugo Genèse, incidences géographiques et écologiques sur la Guadeloupe,* Françoise ed. Pagney Benito Espinal, and Edouard Benito Espinal, 154–161, PNG/Région Guadeloupe.

Breton, Raymond 1656. *Relations de l'île de la Guadeloupe* (éditions originales 1647, 1654 et 1656). Reprinted 1978 by Société d'Histoire de la Guadeloupe, Bibliothèque d'histoire antillaise, Basse-Terre, France. 1665 *Dictionnaire caraïbe-français.* Reprinted 1999 by Karthala Editions Bibliothèque d'histoire antillaise Paris, France. http://horizon.documentation. ird.fr/exldoc/pleins_textes/pleins_textes_7/b_fdi_03_02/010017260.pdf Accessed Jun 15 2015.

Callaghan, R. 2011. The question of the aboriginal use of sails in the Caribbean region. In *Proceedings of the 22th Congress of the International Association for Caribbean Archaeology,* 121–135, Kingston, Jamaica.

Chlieh, M. 2003. *Le cycle sismique décrit avec les données de géodésie spatiale (Interférométrie radar et GPS différentiel): Variations spatio-temporelles des glissements stables et instables sur l'interface de subduction du Nord Chili.* Paris, France: Thèse de l'Institut de Physique du Globe de Paris.

Guadeloupe, D.A.C. 2015. *Archaeological map of Guadeloupe.* DRAC Guadeloupe, Basse-Terre, France: Unpublished data. Service régional de l'archéologie.

De La Porte, Joseph. 1775. *Le voyageur français, ou la connaissance de l'ancien et du nouveau Monde, mis au jour par M. l'Abbé Delaporte (Louis-Abel de Bonafous, Abbé de Fontenay et Louis Domairon).* Cellot, France.

Delpuech, André, 2005. Les « Anthropolithes » de la Guadeloupe. Aux origines de l'archéologie antillaise. In *Proceedings of the 20th International Congress for Caribbean Archaeology,* 443–448, Santo Domingo, Republica Dominicana.

Faujas-Saint-Fond 1804. Correspondance: Géologie. Annales du Museum national d'histoire naturelle. Paris, France.

Fitzpatrick, Scott M. 2010. On the shoals of giants: natural catastrophes and the overall destruction of the Caribbean's archaeological record. *Journal of Coastal Conservation* 14(1): 1–14.

Fitzpatrick, Scott M. 2013. Seafaring Capabilities in the Pre-Columbian Caribbean. *Journal of Maritime Archaeology* 8(1): 101–138.

Gagnepain, Jean, Caroline Luzi, Nathalie Serrand, and Christian Stouvenot, 2007. *Première approche du site de Caille à Bélasse. Campagne de 2006–2007. Commune de la Désirade îlets de Petite Terre et Terre de Bas.* Basse-Terre, France.

Gagnepain, Jean, Caroline Luzi, Nathalie Serrand, and Christian Stouvenot, 2010. Désirade Petite-Terre, Terre-de-Bas, Caille-à-Bélasse. *Bilan scientifique de la région Guadeloupe, des collectivités de Saint-Barthélemy et Saint-Martin.* Service Régional de l'Archéologie, DRAC Guadeloupe, pp. 67–69, Ministère de la Culture et de la Communication, Basse-Terre, France.

Gannier, Odile. 1996. À la découverte d'Indiens navigateurs. *L'Homme* 36(138): 25–63.

Garrabé, François, and Patrick Andreieff. 1988. *Carte géologique du département de la Guadeloupe, 1/50000, Feuille 2 Grande-Terre.* Orléans, France: BRGM.

Graham, E., D.M. Pendergast, J. Calvera, and J. Jardines. 2000. Excavations at Los Buchillones. *Antiquity* 74(284): 0001.

Grouard, Sandrine, Dominique Bonnissent, Patrice Courtaud, Pierrick Fouéré, Arnaud Lenoble, Gérard Richard, Thomas Romon, Nathalie Serrand, and Christian Stouvenot, 2013. Fréquentation amérindienne des cavités des Petites Antilles. In *Procedings of the 24th Congress of the International Association for Caribbean Archaeology*, 277–295, Fort-de-France, Martinique, France.

Guibert, Jean-Sébastien, 2010a. Identification of Anse à la Barque Shipwrecks (Guadeloupe FWI): Historical Research in the Service of Underwater Archaeology. In *ACUA Proceedings 2010*, ed. Chris Horrell and Melanie Damour 157–161, Advisory Council on Underwater Archaeology.

Guibert, Jean-Sébastien, 2010b. Étude historique sur le potentiel patrimonial sous-marin de la rade de Pointe-à-Pitre. Port Autonome de la Guadeloupe/Université des Antilles et de la Guyane, Pointe-à-Pitre, Guadeloupe, France.

Guibert, Jean-Sébastien, 2013. *Mémoire de mer Océan de papiers Naufrage, risque et fait maritime à la Guadeloupe (fin XVII-mi XIXe siècles).* Doctoral dissertation, Université des Antilles et de la Guyane, Guadeloupe, France.

Guibert, Jean-Sébastien, 2014. A Question That Counts in West Indies Maritime Archaeology: Linking Historical and Archaeological Sources. In *ACUA Proceedings 2014,* ed. Charles Dagneau and Karolyn Gauvin 113–120, Advisory Counsil on Underwater Archaeology.

Guibert, Jean-Sébastien, and Franck Bigot, 2014. Entre terre et mer: Infrastructures littorales dans le Grand Cul de Sac Marin de la Guadeloupe. *Archéologie Caraïbe,* ed. Bérard Benoît and Losier Catherine 133–152, Collection Taboui n°2, AIHP/GEODE, Université des Antilles et de la Guyane, Leiden, SideStone Press Academic.

Guibert, Jean-Sébastien, Christian Stouvenot, Fabrice Casagrande, Sandrine Grouard, and Nathalie Serrand. 2013. *D'îles en îles Étude archéologique de la frange sous-marine de l'îlet à Colas (Grand Cul-de-Sac Marin) et de l'îlet à Chasse (Petit Cul-de-Sac Marin).* Petit-Bourg, Guadeloupe, France: AAPA.

Hamilton, Donny L. 2008. Port Royal, Jamaica: Archaeological Past, Present, and Future. *Underwater and Maritime Underwater Archaeology in Latin America and the Caribbean,* ed. Margaret E. Leshikar-Denton and Pilar Lunar, 259–269. California, United States: One World Archaeology, Walnut Creek.

Ingénieurs du Roi. 1768. *Carte des Ingénieurs du Roi,* Bibliothèque Nationale de France, Ge SH 18 pf 155 2 14.

Island Resources Foundation 201. *An Environmental Profile of the Island of Anegada, British Virgin Islands.* Island Resources Foundation, Tortola, British Virgin Islands and Washington, DC. http://www.irf.org/documents/BVI/BVI%20Environmental%20Profile/Anegada%20Environmental%20Profile-FINAL.pdf. Accessed Jun 15, 2015.

Leclerc, F., N. Feuillet, G. Cabioch, C. Deplus, J. Lebrun, S.F. Bazin, F. Beauduce, G. Boudon, A. Lefriant, L. De Min, and D. Melezan. 2014. The Holocene drowned reef of Les Saintes plateau as witness of a long-term tectonic subsidence along the Lesser Antilles volcanic arc in Guadeloupe. *Marine Geology* 355: 115–135.

Martias, Rosemond. 2005. *Rapport de diagnostic archéologique Le Haut du Bourg Saint-Francois.* France: Basse-Terre.

Moreau, Jean-Pierre, 1992. Recherche de vestiges de la flotte de 1603 naufragée en Guadeloupe. *Bilan scientifique du Drassm* p. 65, Direction générale des Patrimoines, Ministère de la Culture, Marseille, France.

Murphy, A. Reg. 2003. Underwater Archaeology in the Nelson's Dockyard, Antigua: The Sea Wall Project. In *Proceedings of the 20th International Congress for Caribbean Archaeology*, Santo Domingo, Republica Dominicana.

Newsom, Lee A., and Elizabeth S. Wing. 2004. *On land and sea Native American uses of biological resources in the West Indies*. Tuscaloosa, United States: University of Alabama Press.

Nicholson, Desmond V. 1976. The Importance of Sea Levels. *Caribbean Archaeology, Journal of the Virgin Islands Archaeological Society* 3: 19–23.

Nicholson, Desmond V. 1977. Pre-Columbian Seafaring Capabilities in the Lesser Antilles, *Compte rendu des communications du sixième congrès international d'étude des civilisations précolombiennes des Petites Antilles*, 98–105, Pointe à Pitre, Guadeloupe, France.

Ramcharan, Shaku Renne, 2004. Caribbean Prehistoric Domestic Architecture: A Study in Spatio-Temporal Dynamics and Acculturation. *Electronic Theses, Treatises and Dissertations* Paper 2133. http://diginole.lib.fsu.edu/etd/2133. Accessed Jun 15, 2015.

Richard, Gérard. 1994. Premier indice d'une occupation précéramique en Guadeloupe continentale. *Journal de la Société des Améicanistes* 80(1): 241–242.

Rodriguez, Miguel, 1992. Excavaciones en Punta Candelero, Puerto Rico: Informe preliminar. In *Proceedings of the 13th International Association Congress for Caribbean Archaeology*, 605–627.

Serra, Laurence, 2012. *Zone de dépotoir portuaire de la Rade de Saint-Pierre (Martinique) Rapport d'opération*. Arkaéos LA3 M, Aix-en-Provence, France.

Siegel, Peter, John G. Jones, Claudette A. Casille, Nicholas P. Dunning, and Deborah M. Pearsall, 2009. Analyse préliminaire de prélèvements sédimentaires en provenance de Marie-Galante. *Bilan scientifique 2006-2008*, Service Régional de l'Archéologie, DRAC Guadeloupe, pp. 139–141, Basse-Terre, France: Ministère de la Culture et de la Communication.

Stouvenot, Christian. 2006. *Rapport de visite à l'Ilet à Fajou Grand Cul-de-Sac*. Basse-Terre, Guadeloupe, France: DRAC Guadeloupe.

Stouvenot, Christian, 2010. *Le dépotoir précolombien du site de Belle Plaine. Une occupation troumassan-troumassoïde. Les Abymes. Guadeloupe Antilles française*. Basse-Terre, Guadeloupe, France.http://hal.archives-ouvertes.fr/hal-00559473/. Accessed Jun 15, 2015.

Texier, Pierre, and Fabrice Casagrande. 2011. *Relevé topographique des ancres et des mouillages présents sur la « Caye » du port du Moule*. Guadeloupe, France: Basse-Terre.

Toscano, M.A., and Æ.I.G. Macintyre. 2003. Corrected western Atlantic sea-level curve for the last 11,000 years based on calibrated 14C dates from Acropora palmata framework and intertidal mangrove peat. *Coral Reefs* 22: 257–270.

Watters, David R, and J. Donahue, 1990. Geoarchaeological Research on Barbuda, Antigua, and Montserrat. In *Proceedings of the 10th International Congress for Caribbean Archaeology*, 375–379, San Juan, Puerto Rico.

Watters, David, 2001. Preliminary report on the correlation of Archaic-Age localities with a paleoshoreline on Barbuda In *Proceedings of the 19th International Congress for Caribbean Archaeology*, pp. 102–109, Aruba.

Weil Accardo, Jennifer, Nathalie Feuillet, Paul Tapponier, Pierre Deschamps, Guy Cabioch, Florence Lecornec, Eric Jacques, John Galetzca, and Jean-Marie Saurel, 2010. Paleoseismology, seismic cycle and tectonic coupling of the Lesser Antilles subduction zone: Insights from micro-atolls. *EGU General Assembly 2010, May 2–7 2010*, 72–97, Vienna, Austria.

Yvon, Tristan. 2011. *Compte-rendu de prospection à l'îlet Kahouanne*. Basse-Terre, Guadeloupe, France: Direction des affaires culturelles de Guadeloupe. Service régional de l'archéologie.

Yvon, Tristan. 2012. Les îlets du Petit-Cul-de-Sac Marin et du Grand-Cul-de-Sac Marin à la Guadeloupe, attrait économique et occupations coloniales aux XVIIIe et XIXe siècles. *Bulletin de la Société d'Histoire de la Guadeloupe* 163: 17–41.

Chapter 10
Conclusions/Discussion

James P. Delgado

Humans live on a dynamic planet that has shaped our lives, our civilizations, and our cultures for thousands of years. Discussions over how we have interacted with our environment and forces such as volcanic eruptions, earthquakes, ice ages, and epic floods have been part of the discussion in archaeological circles for many years. Archaeologists have also paid more attention to cyclical events like typhoons, monsoons, hurricanes, and El Niño and La Niña. As modern technology makes the world "smaller," global archaeology has shown how our species' interaction with the environment is perhaps the most critical element in the human story, especially now when the evidence shows that our species not only responds to the environment, but is now shaping it on a planetary scale.

Increasingly, archaeology has also focused on humans in the marine environment, especially using the lens of maritime cultural landscapes. Over the past decades, Westerdahl's original, Scandinavian concept of the "maritime cultural landscape" has evolved through adaptation and practice throughout the world. That work has included both theoretically driven academic research and pragmatic approaches made in cultural resources management-driven applications. It has also come face to face, especially in the latter circumstance, with the reality that the processes endemic to the "formation", as well as the conceptualization and uses of the maritime cultural landscape, flow into the present and will continue into the future. To define and depict this reality is what the editor sought and has achieved in this volume.

A powerful theme is the illustration of the effects of change at sites over time, and the intersections between the human and natural worlds. The idea that landscapes or sites possess an absolute level of "pristineness" is a never-to-be achieved reality as all things change, sometimes in response to ongoing actions by humans.

J.P. Delgado (✉)
Maritime Heritage, Office of National Marine Sanctuaries, National Oceanic
and Atmospheric Administration, Silver Spring, MD, USA
e-mail: james.delgado@noaa.gov

© Springer International Publishing AG 2017
A. Caporaso (ed.), *Formation Processes of Maritime Archaeological Landscapes,*
When the Land Meets the Sea, DOI 10.1007/978-3-319-48787-8_10

Prehistoric sites on Santa Rosa Island, as Jazwa reminds us, are being impacted by fluvial and eolian erosion as a direct result of nineteenth and twentieth century ranching practices that denuded the landscape. The processes of building on deltaic soils that compact may have helped sink Thonis-Heracleion. At the same time, the end of the last ice age led to sea level rise, which changed the landscape of Thonis-Heracleion, California's Channel Islands, and Guadeloupe, as discussed in this volume, along with much of the rest of the world. A 1700 tsunami, following a 9.0 earthquake, transformed a 1693 shipwreck on the Oregon coast, and in the twentieth century, planting nonnative plants stabilized the dunes that previously had been dynamic elements at the wreck site.

In the introduction, Caporaso argues that archaeology must be studied in conjunction with the landscape in which it exists. This fits well with the ongoing push to move the relatively young field of maritime archaeology beyond a particular shipwreck or group of wrecks, beyond events, and beyond particularism to combine the archaeological and historical record, elements of a maritime cultural landscape, and oceanographic processes to "develop a model that takes into account not only shipwrecks but all archaeological remains in a region" (Caporaso, this volume). Summarizing the current thinking about this new theoretical push, Caporaso draws on the scholarly debate that grew out of Westerdahl's initial posit of a "maritime" cultural landscape (1998), and then focuses on the landscape of shipwreck disasters, drawing into the discussion a review of evolving thinking on site formation processes. Arguing against as O'Shea (2002) has noted, the "illusion of site uniqueness," this volume looks at the question of localized and regional environments, and the question of how the environment affects submerged resources and how they in turn influence the environment.

Horlings and Cook powerfully show this in their examination of Elmina in coastal Ghana. There, on a seemingly inhospitable shore, Europeans established a fortified trading post that utilized the rugged, rocky shore and heavy surf for protection, relying on offshore anchorages and small boats to conduct their business. Modifying features to suit their activities, they were nonetheless occasionally dealt setbacks such as shipwrecks, capsized canoes and boats, and lost cargo and lives. In this dynamic environment as well, an archaeological record survives both offshore and on the shore, which was and remains an access point and zone of exchange. The initial exchanges were for gold and ivory on what was then known as the "Gold Coast," but following the construction of the fort by the Portuguese, the major "commodity" shifted to slaves. Captured by the Dutch in the seventeenth century, Elmina continued to be a major supplier of slaves until 1814. The potential for riches, even when obtained through the worst of businesses, is one underlying behavior in making this "bad" port remain in use for centuries.

Robinson, Goddio, and Fabre exceptionally outline the geomorphologic aspects of their site at the mouth of the Nile and how a marshy, lagoonal environment was ideally situated to be a zone of exchange and transfer, a fortified spot, and a place of government and religious importance. The ongoing processes that made Thonis-Heracleion an important port for centuries also doomed it, as sea level rise, the compaction of water-laden sediments, and tectonic lowering collapsed the port's structures and buried it in sand, mud, and water. The perpetual flow of sediment from the Nile also helped conceal and preserve the site, which in its well-preserved aspect offers much in terms of its infrastructure, occasional "Pompeii"-like micro-sites, and 69 ships. An aspect of the site that bears more examination, they note, is an area of purposely abandoned craft, perhaps not a ship graveyard, but a deliberate attempt to use old ships as either the base for new infrastructure, or to block access to one of the port's basins in time of war.

This raises a fascinating aspect of maritime cultural landscapes—where intensive investment in labor and capital created elements that previously had not existed. What comes to mind is the elaborate waterways of Angkor, crannog settlements, Roman concrete harbors, land filled waterfronts such as eighteenth century New York and nineteenth century San Francisco where the fill included complete ships past their prime, and the modern continuation of "floating cities" of watercraft in Asia, Vietnam's Halong Bay being a prime example.

Guibert, Stouvenot, and Leroy offer a masterful overview of Guadeloupe's maritime landscape and its many sites—that a macroscale view of the landscape offers insights into site placement and changes affecting not only our ability to locate sites, but also site preservation as sea level rise, tectonic uplifting, beach and shoreline accretion, erosion, and tectonic subsidence occurs. The effects of humans in the landscape and on sites are underscored in their discussion of how later colonial period logging may have led to further erosion. Their point is well taken that, while the "sea" was the primary contributing factor in the overall landscape and on individual sites and their formation processes, the role of the sea as a control is poorly understood. Their work in assessing all aspects from storm activity and geomorphology, as well as human actions, bodes well for more work toward the goal of a comprehensive understanding of Guadeloupe's maritime cultural landscape, and by extension, into the larger land or seascape of the Caribbean.

These chapters collectively argue, from a diverse perspective and from different countries, for viewing the maritime landscape as an integrative whole. Landscapes, as Caporaso notes in the introduction, are "maps of human behavior in the long term." A range of activities are reflected in a maritime landscape, as Rönnby (2007) has noted, in economics, coastal settlement and life, the exploitation of marine resources, communication over water, and the mental presence of the sea, with modern aspects such as health, recreation, and aesthetics (including humanistic representations in art, music and literature). Caporaso adds the twentieth and twenty-first century commoditization of the seaward view shed, to which I would specifically note not only include sea and lakeside resorts and condominium developments but also the "gentrification" of former industrial maritime waterfronts in port cities around the world. I would also add another category, that of large-scale conservation through the creation of parks, reserves, sanctuaries, and other marine protected areas.

It is in the "modern" aspects of the maritime landscape where we find a basic aspect of human behavior—that of overlapping and often competing interests and uses. Simply stated,—conflict arises when one group seeks to use and another seeks to control, regardless of motive. Maritime archaeology itself is no stranger to conflict as one aspect of maritime culture, archaeological interest, has clashed with treasure hunting, which without prejudice I will note has been in existence for a much longer period than archaeology. But then, so too has ocean whaling been in existence longer than marine conservation.

Clashes over access to, and use of the maritime landscape withstanding, each and every use reflects human culture and behavior and must be recognized, even if various groups disagree with the use. There are those who decry ocean drilling, and yet the maritime landscape reflects nearly a century of human extraction of energy resources (oil and gas) from the sea. To that end, returning to whaling, humans have hunted marine mammals as a source of energy for millennia. Caporaso recognizes this, and in the introduction cogently point to the truth of landscapes—that they can come to mean many things to many people—through a maritime metaphor when she notes that a site of interest to archaeologists over time can become a natural hazard to shipping, a fish haven as an artificial reef, or a popular locale for recreational diving.

While in the past archaeologists have been aware of the natural and cultural processes by which a once floating ship became a shipwreck, what this volume offers are more examples of the field's evolving beyond individual sites to look at the landscape. It makes sense to go beyond the narrow confines of a single wreck. Ships rarely wreck alone due to a confluence of cultural and natural circumstances. Spurred by market demands, economic necessity, war, or other behavioral drivers, humans often sail ships in dangerous conditions—threading rocky straits, crossing notoriously stormy waters, sailing in thick fog, leaving port with dangerous cargoes, sailing "in harm's way" in time of war, or simply going on a well-known maritime roadstead with bad patches in the road. This aspect of our nature is no surprise, and it is reflected in modern society by cars and trucks on the road, skidding off the highway in snow storms, hitting each other in rainy intersections, or backing into each other in the Whole Foods parking lot. The chapters in this volume all speak well to sites—plural—in the landscape, and again, how the landscape influenced the deposition of the sites within it.

At the same time, they also note another element of this volume—the transformation of sites in the marine environment. There are a variety of what Muckelroy called "filters" at work on sites. As Caporaso states in the introduction, these include chemical reactions with seawater, storm surges, wave action, and biological colonization, as well as cultural processes such as wire-dragging or blasting, contemporary (historic period) salvage, ongoing seabed disturbance through trawling and other bottom fishing, or the removal of artifacts by divers, treasure hunters, or archaeologists. Caporaso, as well as Williams, Marken and Peterson provide detailed essays on how models of site formation, such as Muckelroy's (1978) seminal work, Schiffer's C and N-transforms (1987), Ward et al.'s (1999) natural transformational process, and Gibb's (2006) stages in a shipwreck and finds, while useful and powerful at times, can be rigid, as they do not address material moved off a site that could remain in the archaeological record. Or, as Williams et al. show in their discussion of the "Beeswax wreck" in Oregon, a site can be defined by phases of initial wrecking, tsunami redistribution, salvage by the indigenous population, nineteenth and twentieth century salvage by Euro-American settlers, beach accretion, and modern dune stabilization.

It all illustrates, as Caporaso argues in her chapter, that anthropogenic and natural phenomena cannot be separated and what is needed is a new model in which the total landscape is treated as an "irreducible socio-natural system" as posited by McGlade (1999). This is a landscape studied through multiple analytical arguments with a variety of model scenarios with different temporal and spatial scales. It is, as she notes, a question of continuous cycling, a dynamic environment, and how this relates to "feedback" that drives both continuity and change in human behavior in the maritime environment. I think Caporaso has done a fine job in assessing the still growing body of literature on site formation, merging it with landscape theory, and proposing a more holistic approach that fits well with where we as a field are heading, which is more onsite and post-field work analysis with other disciplines, not just to understand the wrecks, but their place in the larger marine environment, and how it all reflects on what we do as humans on and around the water.

Archaeology has always been about extracting the maximum amount of information from sites which have been subjected to the "ravages" of time and subsequent human activities. Whether it has been carefully extracting millennia-old burials crushed nearly flat beneath the tells of Ur, carbonized furniture from Herculaneum, fields plowed through ancient settlements, sorting through the discarded materials left by tomb robbers, or analyzing grave goods recovered from the black market by Interpol, terrestrial archaeologists have confronted the limitations placed on their research. They work within the reality that time moves forward, as do contemporary societies, with competition for resources and different priorities not always giving the archaeologist a "pristine" site. As a discipline, we always find ourselves at times reevaluating sites based on new techniques, new insights, or correcting the mistakes of pioneering generations in the field—and here I think particularly of Belzoni's battering ram and Schliemann's trench at Troy. What I particularly like about this volume is that within its chapters, there is a focus on learning what one can, with a variety of tools. We do so with the implicit recognition that in this work, what you find is what it is, and we deal with it as best we can as scientists, as scholars, and as human beings who understand our place in space and time.

In the 1980s, it became clear that shipwrecks on beaches did not simply smash into meaningless scatters of splintered wood. Substantial portions of ships survived, even if fractured and distributed over a large area. In some cases, wrecks bedded down into the sand and were buried with contents intact. The famous East Indiaman *Amsterdam* on the beach at Hastings, England, is just one example, but there are hundreds more around the world (Marsden 1975; Delgado 1997a, b:57–59). What has also emerged from work in the Arctic in the 1980s through the last few years is that wrecks of ships caught in ice are also not chewed to pieces and "lost" (Delgado 1997a, b:36–40). What research showed then and now is that understanding the larger landscape, and the oceanographic and geomorphologic processes—of eroding and accreting beaches, and of ice movement in these cases—is not only relevant but key to understanding not only the sites, but the entire region or area and its overall maritime landscape. As one who has and continues to work on sites once thought "marginal" in beach and Arctic landscapes, it is gratifying to see that many of these chapters speak to sites in "dicey" environments where common sense assumptions about sites being "destroyed" or hopelessly scrambled simply do not hold true.

Two chapters have particular significance as they have messages beyond physical approaches to archaeological work. One is Fowler and Rigney's. First, it should be a requisite part of our work that when we work "elsewhere," either in another country or more to the point, in landscapes and on sites where there are indigenous people, culture, and tradition, that permission be sought, and the outside scholar(s) work with the indigenous people as peers. Fowler has done exceptionally well in this regard, and Rigney's knowledge, insights, and sharing of them make this a transformative chapter. Looking at the maritime landscape of the Yorke Peninsula through the lens of Narungga culture and tradition, we see a move beyond the narrative and perspectives of the colonial world. We see an ancient people and their knowledge and beliefs reflected in seamarks and transit lines, fishing grounds, and boat handling. We see a people who practiced both resistance and persistence, adapting resourcefully to western, colonial impositions and changes by practicing in-kind transactions, selling products and their labor, working as fishers, building, owning and operating boats, serving as pilots, and assisting in times of shipwreck.

The other chapter is Wells' examination of the American maritime "frontier" that was the eastern seaboard at the advent of the nineteenth century. It was "dark and desolate" to mariners and to the majority of the population who settled far from the "barren and unforgiving" coast with few exceptions. What drew the attention—and in time developed a maritime cultural landscape in response—were littoral shipwrecks. Wells writes of how the wrecks brought salvagers, speculators, government agents, as well as reporters, authors, and artists who illuminated the landscape as much as helped to define and shape it. This is a fascinating chapter, as it speaks to an environment that changed as the government built lighthouses and other aids to navigation in response, as humanitarian organizations (and in time the government) built life-saving stations, and how the publication of Coast Pilots, based on the work of pioneering government scientists and technicians in the U.S. Coast Survey and the anecdotal knowledge of experienced mariners, was encapsulated in a single volume (and in charts) to provide a guide to the shore. Drawn by the drama, and the pathos, newspapers filled their columns with accounts of wrecks, authors were inspired to offer fictional accounts as well as poems, and artists filled canvases with depictions of desolate shores and shattered hulks. As well, an all-American archetype, the coastal "wrecker" who lures ships to their doom for sake of plunder, emerged in literature as an inhabitant of that maritime landscape.

In the introduction, Caporaso argues that the results of the studies presented here show that the formation of the maritime archaeological landscape (which I would agree is a subset of the larger maritime cultural landscape) and changes in maritime activity are irreducibly diachronically and spatially linked, and the co-evolution of landscape and behavior is evident in the archaeological record. Ultimately, what this book is about is how complex and dynamic the processes by which the landscape and the sites we study came to be. It is about our relationship with the marine environment, arguably the dominant environment on this ocean planet. It is also how we as archaeologists can and should work within the landscape and its environment to extract as much information and insight as we can. This volume shows that we are progressing as a discipline and that the next decade of projects, those within and not reported in this volume, will continue to refine and build on archaeology that seeks to better understand our relationship to the seas that spawned life on earth and continue to influence it.

References

Delgado, James P. (ed.). 1997a. *British Museum Encyclopaedia of Underwater and Maritime Archaeology*. London, England: British Museum Press.

Delgado, James P. 1997b. Beached Shipwrecks. In *British Museum Encyclopaedia of Underwater and Maritime Archaeology*, ed. James P. Delgado. London, England: British Museum Press.

Gibbs, Martin. 2006. Cultural Site Formation Processes in Maritime Archaeology: Disaster Response, Salvage and Muckelroy 30 Years On. *International Journal of Nautical Archaeology* 35(1): 4–19.

McGlade, James. 1999. Archaeology and the Evolution of Cultural Landscapes: Towards an Interdisciplinary Research Agenda. In *The Archaeology and Anthropology of Landscape: Shaping You Landscape*, ed. P.J. Ucko, and R. Layton. London, England: Routledge.

Marsden, Peter. 1975. *The Wreck of the Amsterdam*. New York: Stein and Day.

Muckelroy, Keith. 1978. *Maritime Archaeology*. London, England: Cambridge University Press.

O'Shea, John. 2002. The Archaeology of scattered wreck-sites: Formation Processes and Shallow Water Archaeology in Western Lake Huron. *International Journal of Nautical Archaeology* 31 (2): 211–227.

Rönnby, Johan. 2007. Maritime Durées: Long-Term Structures in a Coastal Landscape. *Journal of Maritime Archaeology* 2: 65–82.

Schiffer, Michael B. 1987. *Formation Processes of the Archaeological Record*. Albuquerque, New Mexico: University of New Mexico Press.

Ward, I.A.K., P. Larcome, and P. Veth. 1999. A New Process-based Model for Wreck Site Formation. *Journal of Archaeological Science* 26: 561–570.

Westerdahl, Christer. 1998. *The Maritime Cultural Landscape: On the Concept of Traditional Zones of Transport Geography*. Institute of Archaeology and Ethnology: University of Copenhagen, Copenhagen, Denmark.

Index

0-9
17th century, 144, 148
18th century, 90, 96, 102, 197–199, 205
19th century, 28, 33–38, 40–44, 47, 48, 141, 142, 148–150, 159, 189, 197–199, 202–204
20th century, 10, 28

A
Africa, 79–85, 87–90, 93, 100, 101, 103
Archaeological synthesis, 211, 212, 214, 216
Archaeology, 1, 3
Australia, 53–57, 59, 60, 63, 67, 68, 70, 72

B
Beeswax Shipwreck, 141, 142, 144, 146–151, 154–156, 158

C
California, 163–165, 167
Cape Cod, 32, 33, 36, 43, 46, 48, 49
Caribbean, 189, 192, 197, 202, 206
Channel Islands, 163, 164, 167, 169, 184
Coastal archaeological site, 163, 165, 167, 174, 180, 181, 183, 184
Coastal Frontier, 34, 38–49
Coastal navigation, 36, 41, 44
Colonialism, 55, 61

E
Egypt, 113, 116, 121, 122, 126, 128, 131, 134, 136, 137
Elmina, 79–84, 87, 89–94, 97–102

E
Erosion, 163–165, 167–169, 172–174, 176–184

F
Formation processes, 1, 4, 5

G
Geomorphology, 190
Ghana, 79–81, 84, 86, 88, 91–94, 99, 100, 102
Guadeloupe, 189–196, 198–200, 202–206

H
Humanitarian aid, 42, 43

I
Indigenous peoples, 53–56, 58, 60, 62, 63, 68–70, 72
Inundation, 125, 132, 135

L
Lake Huron, 7, 21, 28
Landscape formation theory, 10
Landscapes, 1–5, 211, 213, 214, 216

M
Manila galleon, 144, 150
Maritime, 1–5, 211–216
Missions, 53, 55, 56, 71, 72

N
Narungga, 53–55, 57, 58, 60–62, 66, 69, 72, 74
Nile Delta, 113, 114, 117

© Springer International Publishing AG 2017
A. Caporaso (ed.), *Formation Processes of Maritime Archaeological Landscapes,*
When the Land Meets the Sea, DOI 10.1007/978-3-319-48787-8

O

Oregon, 141, 145, 148, 149, 151, 153, 158

P

Paleoindians, 144
Portuguese, 81
Prehistory, 190, 205

S

Shipwrecks, 7, 9–12, 14, 15, 17, 21–28, 32–34,
    37–49, 79–81, 85, 91–93, 100, 101, 189,
    190, 192, 198–202, 204

Social complexity, 217
Spatial analysis, 26

T

Theory, 4, 211, 215
Thonis-Heracleion, 113–137
Thunder Bay, 7, 21
Tsunami, 144, 151–153, 155–157, 159

W

Watercraft abandonment, 120, 121, 125, 130,
    131, 133–135

Printed in the United States
By Bookmasters